TRANSFORMATIONS
7 Roles to Drive Change by Design

BIS Publishers
Building Het Sieraad
Postjesweg 1
1057 DT Amsterdam
The Netherlands
T +31 (0)20 515 02 30
bis@bispublishers.com
www.bispublishers.com

ISBN 978 90 6369 457 9

Designed by Joyce Yee
Illustrated by Kate Ayre (kate_100@live.co.uk)
Copy-edited by Bonnie Broussard

TRANSFORMATIONS
7 Roles to Drive Change by Design

Joyce Yee
Emma Jefferies
Kamil Michlewski

BIS Publishers

Contents

Section 1: Case Studies

Section 2: Expert Interviews

Acknowledgements

We feel very fortunate and honoured to have been given access to organisations courageous enough to share their stories with a wider audience and to lay bare their challenges and struggles so that we can learn from them. Our aim with 'Transformations' has always been to reveal how design is used to enable and support change in organisations. In doing so, we hope it helps others understand, articulate and ultimately apply design in a more impactful way. While we were driven by a professional curiosity to do so, all three of us are further motivated by a personal belief that design can truly help organisations become more human-centric. It is this humanity that we believe has shone throughout these stories. Ultimately design is about people, and the common thread through our book is a testament to this. While we have been keen to focus on the human side of change, it is also important that we offer you the reader a way to enact these changes through the roles we have identified.

This book has truly been a collaborative effort. Not just between us the authors, but also with our contributors from the various organisations involved. They not only allowed us to learn from their experiences, but also acted as 'Friendly Challenger' during the formation and development of our seven roles. We would like to extend a huge thank you to everyone who joined the conversation and generously contributed their stories and experiences. We would also like to personally thank Lauren Tan, who played a big part at the start of our journey in helping us hone the idea into a coherent proposal and is still a huge supporter of the project–thank you! We are also indebted to Kate Ayre, our illustrator, who brought the roles and stories to life with her engaging and intricate drawings. Finally, we would also like to thank Clive Grinyer who has written a perceptive and wonderfully supportive foreword to help set the tone from the start.

We have been very lucky to work with BIS Publishers for our second book. Rudolf, Bionda and the BIS team have been magnificent, giving us the right kind of encouragement while allowing us to organically develop the direction of the book as the stories and insights emerged during the process. Finally, on an individual note we would personally like to thank a few people.

This book has been a continuing journey from the first book that I wrote with Emma and Lauren Tan–'Design Transitions'. Tracking how design has changed has inevitably led us to trying to understand how organisations are changing. Any endeavour, if done right, will always have equal measure of pleasures, challenges and rewards. 'Transformations' was no different. I felt incredibly lucky to have been able to speak to extraordinary individuals leading change in their organisations. It has been a challenge pulling all these various stories together and thankfully I had the crucial support of my co-authors: Emma, with her amazing network and ways to find interesting people for us to talk to; and Kamil who helped shape the book's intellectual spine. This experience has been incredibly intense. Throughout it all, I am so thankful to Ray, for being so patient, supportive and understanding. You have allowed me to focus on the job at hand while keeping me sane. I am also indebted to my parents, Sze Mun and Yoke Sum, from whom I have inherited my stamina and intellectual curiosity. I would also like to take the opportunity to welcome a new addition to the Yee family, my little niece Evynne who alongside Ethan her older brother has shown me how important it is to retain an open and tolerant view of the world that is present in all of us when we are young but often lost when we grow up. Change doesn't mean we have to 'grow up'. It just means we evolve. **Joyce Yee**

Throughout my travels, I have been fortunate to work with and share incredible conversations with people across the globe who are shaping the future of organisations. Talking to such an amazing group of individuals has enabled Joyce, Kamil and myself to constantly reframe our questions–without their support and open minds, this book would not have been possible. Thank you so much for coming on the journey with us. Joyce, your determination and grit continues to astound me every time I work with you and Kamil, you have an amazing way of framing things that are often hidden–thank you for being a dear friend. I am also grateful to the Design School at Northumbria University in the UK, which helped nurture our curiosity about design and where we as authors came together for the first time. My Mam, Dad and brother–Angela, Hugh and Mark Jefferies–without your support, love and encouragement I would not have been brave enough to go on the adventurous path I've taken. **Emma Jefferies**

This book is the next step in a long and fascinating journey for me. What started as an innocuous observation that the design profession is quite unlike any other in the business domain in the year 2000, has now led to one PhD, two books and an unhealthy obsession with the impact of design. As a consultant, I have the pleasure of seeing first hand the transformative power design and designers have. It is my belief that 'Transformations', in the right hands, will be a very potent tool to understand, shape and implement design-driven change. Taking this opportunity, I would like to say a massive thank you to my co-authors, the amazingly tenacious Joyce Yee and the wonderfully connected Emma Jefferies. This thing would not have happened without you. Alex and Mark, my two little boys, have been patient when their daddy was sneaking upstairs to write. I am grateful that they are there for me. Last, but not least, I would like to thank my determined wife, who is also vigorously pursuing big dreams of her own. **Kamil Michlewski**

Foreword

If the 20th Century has been about the story of *design and designers*, the 21st Century has been about the *users of design*.

Following great individual designers that help shaped and form the commercial and cultural success of manufacturing and technology, the 21st Century has been about taking the very elements of design and applying them to new and challenging contexts.

The way that designers have worked, often unconsciously and intuitively, has been unpacked, codified and the component parts shared with a wider audience beyond designers. Being a 'designer' has generally meant someone who has studied design. However it is well understood that, as a rule of thumb, 90% of people who impact on the design of anything often do not have the word 'design' in their job title. The decisions of many, consciously or not, impact and affect the outcome as much if not more than the greatest designer. Where this awareness is conscious, there is a desire to embrace and share the process of design to a wider audience to understand how design can be a more conscious activity in which everyone has a role.

This story is now focused on the influence of design as a thought process and a way of solving problems, spying new opportunities for innovation and for managing ideas and creativity. The impact of design practice on the public sector, commerce and the design of the intangible service experience as well as the tangible and physical, has been readily embraced to the point where we can begin to see design as a thinking process and approach in a historical context.

'Transformations' is as much a historical document of the way design has transformed from a craft-based discipline to one where design is driving and influencing how companies change, bend to meet new challenges or morph into new forms to exploit different opportunities. The stories in this book capture some of the key moments where organisations of all flavour and purpose have used the characteristics of design to create something new and special.

A formally trained designer will know that they share characteristics that can differ from those in the organisations they design for or are part of. So it is not surprising to see that at the heart of this book are character types, seven roles with varying styles and purposes that can bring something new and valuable to an organisation. Each role captures a characteristic and behaviour of designers and explains how these can accelerate or challenge ideas and cultures. These roles offer both designers and non-designers a common language with which to talk about design-driven transformation. They also form a framework to support organisational transformation so that it delivers on its promise.

The pursuit of innovation remains one of the strongest drivers for improvement and competition in the world around us. However innovation and new ideas require change; changes to the approach and way of thinking of both management and workforce; changes to the ethos and methods used; and changes to underlying business models and competitive strategy. In this book the authors have explored design even further and showed how design and these seven roles help that change to start and be sustained successfully over time.

The messages in this book are incredibly important today. Understanding what people have done and the success they have achieved is the most potent way to share, embrace and then carry out change with confidence and trust. None of these journeys were easy, often coming from incredible aspiration, or tough problems that needed new solutions. These organisations were brave, and bought designers and design methods into their processes to make better what they had. The success of that approach is shown in every one of these important and fascinating case studies. In each we can hear the voices of the people who took the risks and forged the successes. They have organised and explained how design helped them change and innovate so that you can to.

This book is perhaps even more important for professionals and executives from other disciplines than it is for designers. It is aimed at everyone who manages and run businesses that spot great opportunities, or realise that there are or will be changes ahead that they need to prepare for in advance. There are many books on management practice available but few about how design can not only shape your product but also shape your organisation for the better. For designers, they can learn how their skills can be applied at a strategic level to support a company to better handle future change.

As designers have embedded and infiltrated deep inside businesses and governments, their characteristics have influenced and made the changes that are captured so brilliantly and accurately by Joyce Yee, Emma Jefferies and Kamil Michlewski. They tell an important story, one that is the future of design and the future of business and organisations too. Whether it is the combination of an innovative and highly creative design group such as Fjord working synergistically with the enormous and culturally opposite management consultancy firm Accenture or the development of patient care in a Thai hospital, we hear the voices of leading design advocates sharing behaviour and culture that allows us to comprehend how they create better outcomes for their users and organisations.

Across innovation and customer focus, organisations big or small, commercial or public service, the stories shine with insight, learning and practical actions that we can understand and take into our own context.

This book is an invaluable resource that contextualises the way in which design and designers help organisations transform, for those who may come from an entirely different functional and professional area. It is a perfect snapshot of the effectiveness of design and, more importantly, the personalities and behaviours required to make every organisation be people focused, innovative, flexible and successful. In a world that is difficult to predict, having the right type of people and behaviour to give you the energy and ability to sustain innovation has never been more important. The message of this book is one of optimism and innovation, where we all play our role in design, and design what we do to thrive for the benefit of all.

Clive Grinyer
Premier Design Director, Barclays
December, 2016

Introduction

What is change?

Change is about the challenge of moving from one state to another. Naturally occurring change seems effortless; water constantly changes from liquid to gas and to solid as a result environment's temperature fluctuations. Constant flux is the natural state of the physical world surrounding us. Human organisations, in contrast, once created have a strong preference for preserving the status quo. They are, of course, much more complex phenomena, consisting of individual people each with their own dreams, plans, biases and fears.

It is rather ironic that the most adaptable animal on the planet has such a strong preference towards stability as observed from the group level. We are often so obsessed with preserving the status quo that we are willingly endangering our own survival. As Chris Argyris famously observed, we are only prepared to change when our fear of the consequence of failure exceeds that of the unpleasantness associated with the fear of change. Never has it been more important to grasp just how we should best prepare for the hyper-dynamic reality we live in today. We believe that any assistance that we the authors can offer to managers and executives, which leads to more effective and enjoyable organisational change efforts, is something worth pursuing. In this book, we want to shed light on how change is instigated and delivered by designers and design-inspired executives. As we will show through our case studies and expert opinions, there are certain aspects of the design-informed approach and professional culture that are highly effective in pursuing meaningful and effective organisational transformations.

Why change?

Many organisational change efforts are driven by the need to cut costs, reduce complexities or optimise operations. Consultancies are wheeled in and the transformation is clearly labelled and conceived as an explicit and often wide-ranging programme. In contrast, the change we witnessed through our case studies is subtler, yet quite powerful when it comes to delivering value to consumers and users. Companies we spoke to say they want to be more human-centric, they want to be more nimble, they want to deliver better service and they want to be more innovative. Rarely do they speak directly about organisational transformation, yet this is exactly what is happening to many of them in the process.

The common narrative is this: We're not happy with how we've been delivering products or services. We want to create things that people want to use but also meet our business needs. So, we need to understand our users better. In order to understand our users better, we need to involve design.

Once we have designed the new services we wonder how we are going to deliver them based on the existing organisational setup. We then realise that the current system is inadequate on a number of levels. We then grasp that the best course of action is to step back and change the fabric of our organisation. We realised that we can't just change our structure and processes but most importantly we need to help our people manage change by placing them at the heart of the transformation.

This strikes us as a genuinely better way of changing organisations. It's more purposeful and less cynical. Not focussing on the fact that the initial goal is to change the organisation, but instead considering how we are creating better value for the customer and employees, is a more grounded, palatable and transparent approach to organisational transformation. Design and designers do this exceedingly well. Hence, the transformation by design that we've witnessed is the result of a shift in emphasis towards the people the organisation serves (both internally and externally), rather than the structures the organisation possesses. As such, it could be seen as more natural, less confrontational way to enact change.

Why use design for change?

We acknowledge that design is not the only change agent enacted for organisational transformations but is one of the more effective ones if we are aiming for a more human, engaging and resilient process.

It's often more meaningful to focus on individuals and understanding their personal stories, rather than focusing on many people and hunting for patterns of behaviour. This way of operating gives the design approach an advantage over more traditional approaches. It helps to distil powerful, actionable human truths, which enrich not only the user stories, but ultimately lead to better products and services. In contrast to the Learning Organisation school of thought, which suggests focusing on systems thinking as a means of changing people's mental models, design-centric approach offers human-centricity, deep empathy and the persistent concentration on the needs of not just the organisation's customers but also its employees.

Moreover, other professionals in the organisational domain see designers in a relatively benign light. This is largely due to their seemingly apolitical stance and a primary focus on the needs of the users. They are, for the most part, ideologically and politically transparent when it comes to their efforts. This gives them the edge over some change initiatives that are driven by, for example, cost-cutting or efficiency agendas. The design-led approach focuses first and foremost on increased consumer satisfaction and user value. This very fact has an effect on how the transformation instigated and enacted by design is seen by the stakeholders. What at first glance might appear to be a weakness: designers seeming lack of concern for internal politics, apparently weak (albeit certainly growing) professional stature, lack of 'corporate aggression', could in fact help ensure the effectiveness of this approach when it comes to influencing change in organisations. The soft power of this philosophy is often irresistible to employees who see it as genuine, authentic and impartial. As clichéd as it sounds, designers and design-inspired executives are genuinely interested in making the world a better place. This is felt and appreciated by those who work with them in organisations, subsequently giving them a better handle on how to help people change their ways.

How do we use design to drive change?
Seven change roles

In order to help structure the *how* of organisational transformation, we've identified seven crucial roles that design, designers and design-thinkers play in the process. Those roles came into being as a result of the synthesis of the case studies, expert interviews as well as our own extensive experience. We've subsequently used the roles to inform the flow of this book. These change roles are not mutually exclusive and can be activated simultaneously at various points of an organisation's transformation journey, depending on the context. Roles change and can be discarded when no longer needed. Our aim is to provide executives and consultants, who are about to or who have already embarked on a change process, with a framework that can assist them with decoding and enacting change efforts where design is a significant component. We explore the roles in more detail in the next chapter, but here is a brief description of them:

Cultural Catalyst – Design in this role kick-starts a more creative and explorative culture by offering a 'safe space' to try out new ideas and encourage a proactive and experimental attitude.

Framework Maker – Design offers an understandable, coherent and human-centred process and structure to achieve product/service innovation.

Humaniser – Humanising the work and tapping into the deep empathy between the teams as well as with consumers/users.

Power Broker – Design engages people from different parts of the organisation and, by using key user needs as a reference point, reconciles silo-induced interpretations and tensions.

Friendly Challenger – Design takes on the role of a 'critical friend', challenging how things are done in the organisation and in some cases in professions and culture.

Technology Enabler – Uses a design approach to focus on user needs when implementing a technologically-driven challenge.

Community Builder – Uses design to engage stakeholders and help create a culture of participation and involvement that can survive through changing contexts.

3 levels of organisational transformation by design

The Danish Design Ladder[1] is a useful concept for assessing where any given company, or indeed country, is with regards to its implementation of design. We suggest that *Transformation by Design Ladder* can be used to identify and guide organisational change efforts that are informed by design. It includes 3 levels:

Level 1 – Changing products and services

This level is about providing better, more human-centred products and services and improving their value to end users. Here, the transformation by design focuses on the intrinsic nature of what is being offered. Taking the case of Bumrungrad Hospital, which we will discuss at length later, the transformation was to improve patient journeys, which led to a better service for patients.

Level 2 – Changing the organisation

Level 2 transformation takes on the inner-workings of the organisation itself. In this instance, design acts as a lever through the interventions of senior designers and design thinkers, and alters some aspects of organisational structure, strategy, processes or culture. In the case of Spotify, the design function was rebalanced with the dominant tech and product aspects of the organisation to bring a more human focus to their product development.

Level 3 – Changing the nature of organisational transformation

Finally, Level 3 transformation tackles the real obstacles to organisational change such as cultural biases, misaligned vocabulary and points of view, vested vertical structures and many more. Design in this instance is helping overcome and mitigate the natural systemic and social resistance to change as well helping to equip employees with the right tools to carry out the human-centred change. This can be seen in how the US Department of Veterans Affairs helped staff overcome their long-held assumptions and reconnect them with veterans' current needs.

What are the strategies to drive change by design?

Assuming you are a leader in an organisation charged with bringing about change infused with and informed by design, how can you make sure that what you're being asked to do gains traction?

Understanding how design fits in the organisation

From our study of the cases in this book and our experiences beyond, we see that one important consideration helps dramatically in this task. How design-driven intervention is perceived is as significant as what it actually delivers in terms of new processes, new tools and new products and services. Amongst traditional corporate structures, there is a tendency to treat design with suspicion. It is often seen as fluff, difficult to quantify and difficult to explain. In addition, the nature of a design-centric approach is such that it requires a certain level of acceptance of ambiguity and unorthodox methods of obtaining evidence, insights and influence. Therefore, it is crucial that the story, narrative, language and symbols associated with transformation by design are carefully orchestrated to suit company's particular circumstances. Different organisations have found different ways to do this. Telstra, Itaú Bank, Deloitte and Bumrungrad Hospital, for example, prioritise showing value linked to clear KPIs to demonstrate this value immediately to the executive team. At the same time, creating a design function that traverses other functions ensures that its not seen as a 'resource competitor' to the other areas but instead as a resource to draw from and support others.

Aligning design with organisational goals and resources

Having an eye on how the money is spent and where the pockets of power reside also helps in bringing about meaningful change. Say your organisation has committed resources and effort towards a digital transformation. It is not that difficult to make a case that, in order for the digital transformation to be successful, there needs to be a significant design component attached to it. This is how it was done at Adur and Worthing Councils. A Design and Digital Team was established, thus linking the two elements. Their subsequent efforts combined digital delivery and tools with the philosophy and principles of design. Having something concrete to work with, like the suite of digital tools, gave them a strong anchor point and fulfilled an important organisational goal.

Labels matter

It is possible that not labelling design-driven transformational efforts as 'change management' may actually serve to create more trust and goodwill and subsequently prove to be more effective. Big organisational change with capital 'C' and the organisational change carried out as a by-product of people passionately, and yes, often wholeheartedly advocating users' needs, are two distinctively different propositions. The latter is a more positive, more people-centric way to change organisations. It doesn't normally follow the onset of a crisis. Instead, design-driven transformation is a more long-term undertaking, which, at its heart, is the continued effort to make organisation's offer more user-centric and valuable. It also has the capacity to instil a culture of continuous improvement and learning, which is anchored in the customers' needs.

Executive support

This is an obvious point, but without executive level support and understanding of what design is, it's going to be a challenge to drive change by design. Having visionary leaders–like Secretary Robert McDonald (US Department of Veterans Affairs), Jim Hackett (Steelcase) and Hasso Plattner (SAP)–who believe in design's role as a transformative agent, enables long-term resourcing and focus for this purpose. Short of cloning these individuals, it's often necessary to make a strong case for design through business examples, exposure to design value through first-hand experience and bringing in external advocates to raise awareness and interest in design.

How are organisations using design as a change tool?

Along our journey, we've talked at length with a number of companies that have design as a key strategic component of their transformation efforts. These examples illustrate various ways in which transformations happen and the routes it follows. It's also important to note that the stories captured are only a snapshot in time and the change effort continues. However, its been eye-opening to see just how robust and extensive the use of design is in many of them:

At **Steelcase**, the world's largest maker of office furniture was already familiar with and enthusiastic about design, design thinking was able to play a more strategic role in helping to navigate the multitude of cultures that the company was starting to serve. Not only was it a market differentiator for Steelcase, it has influenced they way they operate and most importantly how they manage continuous change in the organisation.

Accenture's acquisition of **Fjord** (a design consultancy) is a typical way in which a large organisation attempts to boost its design capability. In this case the two companies have recognised that respecting each other's qualities and cultures is the best means of generating value. One of the key impacts Fjord has had includes the complete redesign of Accenture's performance management process for their 360,000 employees.

ZOOM Education for Life is the classic case of a change narrative that we have heard in many of our examples. They wanted to revolutionise their product development process to help them exploit the challenges and opportunities in the Brazilian education sector. They created and tested a design driven innovation process, which not only resulted in increased profits but also helped them improve the inner workings of the organisation and overcome professional snobbery in relation to user feedback.

Deloitte Australia had the vision to use design to initiate its own transformation into an innovative, human-centred professional services company. Through a combination of serendipitous events–executives independently meeting leading Design Thinking exponents and a systematic programme, 'Different by Design', they have been able to make significant strides towards being truly design-led.

At **Bumrungrad Hospital** in Bangkok, design was used to improve the patient's overall journey. It's well understood what that entails; hence the design intervention has a clear and unambiguous purpose. This, alongside a clearly sandboxed approach, with two-month trials, helped to create traction for the design-driven paradigm of change.

Design was instrumental in helping **Spotify**, the world leader in online music streaming service, to streamline its innovation processes and ensured that everybody in the organisation spoke the same language. It played a crucial role in helping functions collaborate effectively to bring about change and user-centric service improvements.

Itaú Bank challenged itself to become more innovative and responsive using design in order to meet changing demands in the Latin American banking sector. The result was an explicit innovation strategy and a wide-ranging set of initiatives around three stages co-developed with IDEO called 'Learn', 'Build' and 'Dream'.

Telstra is the market leader in Australia's telecommunications sector. They have positioned design as an approach to help the company become more customer-centric. It aligned their work and concerns around key user needs and away from internal, organisational struggles.

The **US Department of Veterans Affairs**, the 2nd largest department in the U.S. Federal government was looking to turn around their declining service quality and long waiting times for access to VA healthcare. At a critical junction, design involvement through explicit use of customer journeys and personas has been able to markedly raise service quality and trust in the entire institution.

At **SAP** innovation and design thinking have had a longstanding presence. In the latest wave of transformation the Design and Co-Innovation Centre team has been set up to further elevate the role of user experience and design. Within a framework of Advise, Innovate, Empower and Realise the company has been making significant progress at implementing another level of meaningful and human-centred change at an organisational level but more importantly at an individual and personal level.

Adur and Worthing Councils is a local government organisation looking to transform their IT systems into a digital service that not only reduces cost and improves efficiency (organisational benefit) but also transforms the relationship between the council and citizens (human benefit). Design, in effect, was there to make sure the right kind of technology was being used in the right way. It's also an example of where digital transformation often leads to organisational transformation.

Satellite Applications Catapult is pioneering the use of design at a strategic level in the space sector, an industry that is predominantly engineering-led. Here, design has been able to fulfil a number of roles including unlocking the highly technical nature of the endeavour for a sector that is moving to a mass market.

Innovation Studio Fukuoka is a great example of how design-derived impetus can help create a network of projects within a community. This city-sponsored innovation platform demonstrates how design can impact positively on even the most conservative culture, by bringing people together around a humanistic action framework.

How is this book structured?

The book begins by introducing the seven roles to drive change by design. This provides a framework for readers as they read through the 13 case studies to help decode and understand how change was enacted with the help of design. The case studies illustrate how design has been used to drive change by design and they are generally grouped around key change roles. We signposted these roles throughout the case studies to help readers make sense of the 'how'. We also highlight key learning and insights from the case studies, particularly around conditions for impact, challenges, stages of transformations and their motivation for change. We further ground our findings with insights from seven experts from the areas of organisational change, leadership, social innovation and digital transformation.

How do you use this book?

We know that design as a philosophy, a set of methods, and a culture has a growing impact on organisations across industries and sectors. Our seven change roles and the Transformation by Design Ladder are one of many resources that can guide you through this change process. Our aim for this book is to shed new light on how design helps to transform organisations by revealing how it's actually done through the people leading it as well as those experiencing the change. We are witnessing an increasing number of organisations using design to drive change to help them become more innovative, human-centred and resilient. It is our hope that through our book you will prepare yourself better for the ambitious and worthwhile mission of transforming your organisation using design.

More information on specific tools and methods that utilise our findings can be found on the website www.transformations-by-design.co.uk.

Notes
1. The Design Ladder was developed by the Danish Design Centre (DDC) in 2003 as a tool to measure the level of design activity in Danish businesses.

7 Roles of Design

What are the roles and how to use them?

Throughout this book, we consciously use certain shorthand. When we say design 'does' or design 'has' we mean the people who are informed and inspired by the methods, philosophies or values associated with the design paradigm in organisation studies[1]. The same applies to our seven roles. Just as we take on different roles in life–we are a colleague or mentor at work, mother or wife at home, training buddy in the gym, an old friend in a pub–so do we have the capacity to take on different roles when it comes to design in an organisational setting. What roles design does take on, depends on particular circumstances. A collection of 13 case studies in this book allowed us to trace those different circumstances and glean from them some common, underlying patterns. We've interpreted those patterns as the roles played by design professionals and design thinkers involved in influencing the course of a multitude of organisational transformations. It's also important to state that we consider them change roles that use design, rather than roles for designers. Our case studies demonstrate that you do not need to be professionally trained in design to take on these roles.

So, what do we mean by a role? According to the Oxford English Dictionary, a role is defined as 'the function assumed or part played by a person or thing in a particular situation'. In our case, the function relates to the dominant mode in which design acts on the organisational system undergoing change, i.e. are we predominantly concerned with structures and processes or are we zoning in on people, politics or culture? As we've seen, the actors fluidly move from one context to another, bringing with them the right tools, methods and approaches. Say, an organisation is struggling to align disparate functions in order to provide more innovative services to its customers; design (through design professionals and design thinkers) would then assume a role, which is most appropriate given the challenge. It could, for example, intervene to reconcile misaligned groups or act as creative catalyst to encourage more enthusiastic and productive collaboration.

In our quest to analyse how design influences and mitigates organisational change, we've uncovered seven distinct roles (or capacities) to guide our discourse. Together, they form a framework that informs how designers and design thinkers help organisations implement the design paradigm and lead wider organisational change initiatives. These roles are not discrete. They can overlap and complement one another. Some roles have more affinity with other roles and are often used as additional support. The seven roles can be at play concurrently, depending on what is required at any given moment. We also see our roles contributing to growing accounts and frameworks of using design thinking at the organisational change level[2].

In the following pages, we'll expand on what we believe the core essence of each of the role is, together with illustrative examples taken from our case studies. Before we do, however, here is a mini overview, which uses the roles as an assessment framework. The following questions should give you a sense as to whether using design, in the way its described in this book, could bring tangible benefits to your organisation.

Cultural Catalyst
- Is your culture in a state of positive flow or does it feel stuck? (Being 'stuck' could mean: lack of ideas, too many similar ideas, pace of change is too slow, widespread groupthink, rehashing old arguments etc.)
- Does your culture suffer from a lack of vibrancy, colour and creativity or does it posses these attributes in abundance?
- Is there a widespread willingness to share ideas and work on them iteratively together in a spirit of positive and open-minded collaboration?

Framework Maker
- Are your innovation initiatives properly grounded: do they generate too few or too many ideas; are the new products aligned with company's vision and do they genuinely make a difference?
- Do you believe your organisational processes are sufficiently anchored in the real needs and desires of your customers?
- Has your organisation implemented design as strategic mindset that goes beyond styling?

Humaniser
- Do your employees feel empowered or disempowered by the ongoing change process?
- Do you see yourself as in-touch or out-of-touch in relation to the nuanced and ever-changing needs of your users?
- Do you feel like you are telling an engaging and persuasive enough story to your employees and stakeholders in order to influence their behaviour?

Power Broker
- Are you finding it easy or difficult to reconcile multiple voices and specialisms for the benefit of creating a more innovative and robust organisation?
- Do you have a sense that the right or the wrong business function is in charge of the innovation processes?
- Do you feel that there is a strong user voice and a reference point guiding project trade-offs and key investment decisions?

Friendly Challenger

- Does your organisation embrace internal challenge in an attempt to ultimately focus on creating better products and services for the customers?
- Are your functions more interested in defending their turf or serving the needs of the users?
- Do you have effective mechanisms, processes and people to diffuse the tensions inherent in company-wide transformation?

Technology Enabler

- Does technology, including digital technology, work intuitively and is it appreciated as approachable and useful by the employees and external partners?
- Do your digital tools help or hinder human connection with your customers?
- Do you have the right set of technologies to support your innovation process?

Community Builder

- Do you have a habit of creating platforms where a mix of people from different sectors and with vastly different skill sets come together to bring positive change?
- Do you know how to direct a collection of stakeholders towards a common goal anchored in human-centred needs?
- Can you offer tools, such as co-design methods, to enable and encourage community participation and ownership?

Notes

1. There is a growing body of literature, which discusses this topic at length. Design thinking, service design, professional design culture are all strands of this debate. Here we pragmatically assume a certain uniformity of the concept in order to advance the dialogue on how the people who espouse the design-informed paradigm influence organisational transformation.
2. See for example, The Ten Faces of Innovation (2005) by Tom Kelley, The Design of Business (2009) by Roger Martin, Change by Design (2009) by Tim Brown, Serial Innovators (2012) by Abbie Griffin, Raymond Price and Bruce Vojack and Design Attitude (2015) by our co-author, Kamil Michlewski.

Cultural Catalyst

'The culture changes in the most important point for us because in education we have a huge challenge to help our team think differently from the way that they are used to thinking.'
Victor Barros, ZOOM Education for Life

Summary of the role's capabilities

- Stimulating cultures to change through a clear focus on peoples' needs and deep empathy as a means of approaching sensitive, cultural challenges.
- Infusing the culture with the value system based on transparency, continuous feedback loop with the users and attention to cultural, social and individual nuances.
- Promoting openness and pragmatism.
- Looking at the totality of the human experience and making the case for rich cultural interactions.
- Embracing plurality and multiplicity of voices as a core belief, thus creating an atmosphere of trust and welcoming dissenting voices.
- Seeing heterogeneity of ideas and value systems as a springboard for innovation not an insurmountable stumbling block.
- Breaking down internal silos and introducing horizontally-integrated teams.

Changing demographics and changing expectations of what a job is, shapes what's best for the organisational change process. Sharing the values and espousing the purpose of the organisation is paramount for the millennial generation. The closeness of the design profession to actual human needs is very appealing to a generation that is less materialistic and more in-tune with the sense of corporate purpose. Millennials are also more focused on the human beings in the value equation and are more eager to engage in causes with a strong purpose. For these reasons, they are more likely to engage in organisational change efforts that have these components.

The type and ethos of change promoted by the design professionals sits well with Millennials. If the main thrust of the activities is clearly focused on creating the best possible fit with users' needs, then, as the Bumrungrad Hospital case would suggest, the involvement and buy-in from the millennial staff cohort becomes easier. The primacy of the patients' experience serves as a powerful reminder of why the changes are taking place in the first place. The young staff members respond to this sort of purpose-focused transformation more favourably. The fact that the focal point of organisational change isn't on the organisational change itself helps in making it more effective. Past efforts of instigating and successfully carrying out major organisational transformations suggest that focusing on a big and meaningful purpose, which is difficult to argue against, pays dividends. This is yet another reason why design-driven transformation might be an attractive and effective way of bringing along an entire cohort of employees who no longer rely on the role of the authority in guiding their behaviour.

In its role as a *Cultural Catalyst* design has the capacity to augment aspects of national cultures. In the case of Innovation Studio Fukuoka, it helped challenge the patriarchal, top-down societal norms of Japanese culture that often rely on the government to lead and direct. The 3/11 (earthquake) event has proven to be a major turning point in changing the Japanese psyche and has resulted in a rise in bottom-up approaches led by concerned citizens working together to solve community problems. In Fukuoka, design catalysed the community by equipping it with a design thinking approach. The design-led process (Uncover, Inspire, Exchange) brings out the empathetic qualities needed to view issues and challenges through a more humanistic lens. It helped to overcome challenges associated with a rigid set of cultural norms that were not suited to collaborative, venture-focused and transparent efforts. Again, by putting the users at the centre of the action and wrapping it up in an easy to articulate and understandable process, designers and design-inspired leaders were able to shift the debate towards something that mattered.

Another benefit of this role in a cultural context was how it helped overcome the tendency to keep innovation secretive and insular. The standard operating model of closed-door, company-led innovation was challenged successfully not only in the Fukuoka context, but also at Itaú Bank where they are keen to build partnerships with other organisations to deliver a more holistic experience for their customers. A more open, innovative culture encourages building ideas together and trialling them openly and quickly.

At Steelcase, a company with a long tradition of design-centricity–not least through its ties to IDEO–a *Cultural Catalyst* role has come to the fore. Here, after years of working with the design thinking paradigm after its introduction to the company in 1997, the challenge was how to re-enthuse the culture of creativity and user focus. The company has responded by taking design to another level and weaving it into the very fabric of the change process itself. In the early days, they were very much using design as a *Cultural Catalyst* to help the company evolve from it's manufacturing roots to one that is more idea based. Through the creation of an internal WorkSpace Futures unit, Steelcase reaffirmed the central, strategic role of design.

A Brazilian company, ZOOM Education for Life, is an example where the *Cultural Catalyst* role went beyond the organisation itself and is catalysing changes in the wider sector. Through a wide-ranging and successful challenge to the entrenched educational practices, out-dated curricula and professional snobbery, ZOOM was able to deliver meaningful change. Here design, with the help of the *Friendly Challenger* role, was able to modify the behaviour of the educators, by encouraging them to collaborate with students, parents and other partners. Instead of insisting that theirs is the only legitimate way, teaching processionals engaged in a collaborative process that positively influenced the final outcomes.

Framework Maker

'The initial goal of the Design Practice was to demonstrate the value of design and to get the team up and running and settled. And we have achieved that. We have proven that design works.'
Cecilia Hill, Telstra

Summary of the role's capabilities

- Providing an important sociological and psychological safety net for those in the organisation seeking to engage in an exploratory, creative and divergent mode of thinking and acting.
- Making sure the aims and objectives are focused on the paramount importance of the value generated for the benefit of the consumers.
- Propelling the organisation towards a position where it can take full advantage of the opportunities emerging in the fast-changing commercial environment.
- Offering visualised and tangible markers of progress and prototypes, which help to create a pragmatic, purposeful conversation, which in turn drives the human-centred initiatives forward.

Innovation, consumer and user-centricity can often feel like a leap in the dark. It is especially true for those who have been developing their ideas at arm's length from the people who are actually using their products and services. The fear of the unknown can be overwhelming for many individuals. As we write elsewhere[1] designers are comfortable embracing the ambiguity and uncertainty that is inherent in trying to come up with something new, original and useful. Unfortunately, many other professionals are not so enthusiastic about letting go of the illusion of certainty. Moreover, at an organisational level, where accountability and certain predictability are expected as a given, this lack of clear markers about how and in what direction things are developing, can be quite unsettling. This is acutely visible during the attempts to transform the way an organisation works and how it is structured. The *Framework Maker* role that design plays, is key in ensuring the fear of the unknown doesn't paralyse the people involved. What is visible in the change efforts is that establishing a set of design processes and methods, helps to create a framework, which provides an understandable guide and a common language to all involved. It gives the organisation the confidence that it can achieve its goals and ambitions, despite the fact that the precise nature of the outcome is not known at the start of the process.

At Telstra, the ambition was to become a customer-centric organisation and continue to meet the changing demands of the telecommunications sector in Australia. Design is being used to support this through three services offered by Design Practice, an internal Telstra team. Those three are: 'Design-led Strategy', which focuses on facilitated Design Thinking for strategy, 'Customer-led Design & Testing', which helps teams design products and services, and 'Design Capability', which helps embed design in Telstra's culture and practices.

Helping to overcome fear of the unknown and uncertainty by providing a process as a reference point is self-evidently useful. In many cases it is hard to anticipate the exact outcome of many of the change efforts. This fact is difficult to accept by some people in an organisational setting. It's quite a natural reaction to the perceived threat coming from a new set of circumstances. Employees and stakeholders want to be reassured that resources will not be wasted and that the ultimate results will be worth it. The paradox here is that, in order to come up with something that surprises and delights the customer, the team needs to surprise itself in the process. Designers and design thinkers know this and embrace it. With the help of the *Framework Maker* some useful structures are put in place to reassure and to signal how the ultimate goal will be achieved. Design-inspired intervention offers a number of helpful models that help to frame the process. From the classic, IDEO-inspired 'Desirability, Feasibility, Usability', through Telstra's 4Ds: 'Determine, Discover, Design and Test and Deliver', to Itaú Bank's 'Identify,

Develop, Implement'. What all these strategies have in common is that they have been adapted to suit their own organisational context, language and way of working.

In a professional service company Deloitte Australia, design has been playing a central role in changing the overarching nature of their organisational culture. From the serendipitous encounters of Deloitte's executives with leading proponents of the design thinking paradigm in business (Robert Verganti and Roger Martin), to more structured and systematic efforts, the company has been on a path towards becoming design-centric at its heart. The *Framework Maker* role was instrumental in establishing organisational parameters that enabled the successful integration of what was, for this type of business, an unorthodox philosophy. Some of the elements of the design framework at Deloitte Australia included the role of the Design Leader (Maureen Thurston), the launch of the 'Different by Design' programme, establishing the Design for Business team and setting up the Strategy Capability team responsible for the development of the design capability across the entire firm.

Design not only provides a process but also methods that help drive human-centred initiatives. The consumer journey map kills two birds with one stone, as it illuminates the user needs and at the same time empowers the staff to propose their own ideas and improvements, thus engaging them in the change process. The ownership of the transformation can then be neatly transferred to the members of the right team. It boosts people's motivation to follow up and persevere with the effort. What helps significantly is the way the process is instantly made tangible and accessible to all. It doesn't sit in a report, but rather it's visible and can be acted upon and owned immediately.

The simple interactions often work best. At the US Department of Veterans Affairs, using user journey maps and personas gave employees direct contact with users and galvanised people to act. It became an important rallying point for employees and helped overcome previous misconceptions and resistance to change.

Notes
1. See Michlewski K. (2015) Design Attitude.

Humaniser

'I see higher motivation amongst colleagues. I see many examples of people working together and there is a better working environment. Ultimately all organisations are made up of people in the end, therefore the people have to change if you want to change the organisation.'
Jochen Guertler, SAP

Summary of the role's capabilities

- Injecting empathy into the process, creating a human dimension to the work and making business challenges easier to relate to and engage in.
- Bringing personas, journeys, role-play, in-depth explorations, and many other techniques into play to put a human face on the often dehumanised business discourse.
- Creating organisational traction by creating stories and visualisations that inspire people to take action.
- Challenging organisational structures, processes and protocols to understand the customer's experience.
- Offering a dose of humility into the organisational value equation.
- Creating an approachable and inspiring change narrative centred around the purpose of serving real human needs.

Injecting a high dose of empathy and cultural sensitivity into the innovation process is one of the most notable contributions of design. Peeling away the organisational structures, process and protocols to understand customer experience is a powerful way to highlight key issues and recalibrate the organisation's goals towards adding value where it really matters.

There is often a tendency in businesses and organisations to drift away from the people they serve. Young, fledgling institutions and ventures are naturally close to their customers. A single shopkeeper, a craftsman or a start-up business owner comes into direct, daily contact with consumers. They understand their needs and wishes almost viscerally. This is how the original value is created after all, by closely catering to the nuanced needs of real people. With growing size and complexity, organisations can sometimes find themselves several steps removed from the users of their products. It's a natural consequence of the shifting focus towards internal tasks and away from external, organisational adaptation. Something that was once obvious and natural, that all the members of the organisation are ultimately responsible for delivering exceedingly great value to the users, becomes vague and opaque. The responsibility for responding to consumers' needs is delegated to the Marketing and Sales functions, if such exist. In our experience, this distancing from users can have profound consequences on the organisation. There comes a point when the organisation realises that it's time to reconnect with the users and re-learn how to truly add value to their lives. This realisation was what made Hasso Plattner, one of the co-founders of SAP, bring in design thinking in an attempt to close this gap.

Design-led transformations offer a particular brand of this approach, one that relies heavily on the attitudes and values espoused by the professionals forming the core of those design-led teams. This style typically emphasises embracing deep empathy and a quest for the most profound insights linked to the intricate and often hidden needs of the users. SAP offers compelling evidence of how design in the role of a *Humaniser* influences organisations. The case touches on a powerful, personal story of a software engineer who not only felt the power of the user through direct contact but also observed growing empathy between colleagues. In it we learn that, for the first time in his 10-year career, the engineer 'understood the reasons certain software features had to be implemented'. Close proximity to clients and users, tightly knit project teams, early prototyping and a hefty dose of rich communication– all aspects closely linked with the design approach–make a significant contribution. Thanks to design, something that once was a transactional and technical 'IT project' becomes a platform of shared understanding and a means of addressing real user needs.

In another example, Satellite Applications Catapult is using design as a humanising role in a sector that is traditionally very engineering focused.

Their extensive use of designers, including product, interaction, architecture, graphic, UI/UX and service design means the company possesses critical mass to look at innovation and change through the lenses of a human-centred values set. This manifests itself extensively throughout the organisation. One of the results has been the creation of an environment that fosters a strategic and long-term role for design. It has resulted in internal shifts such as improved communications between team members around the user requirements and external initiatives such as the 'Satellites for Everyone' campaign, which lays the foundations for broadening the footprint of the business.

From the US Department of Veterans Affairs we learn that 'great customer experience requires great employee experience'. In order to build a more rewarding staff experience design offers tools to understand users better. At Spotify, design in its guise as a humanising force, has given users a voice in their product development. It particularly helped the tech and product members of their product team to focus on the user and work collaboratively to solve problems.

Another good example of the *Humaniser* role at play is Fjord. Accenture, a global professional consultancy, acquired Fjord in 2013. Accenture's aim was to use Fjord (its people, practices, cultures etc.) to create a more creative and responsive organisation. In return, Accenture provided them with access to cutting-edge technology, data analytics and business and industries capabilities. A significant contribution credited to the design consultancy, which could be classified under the *Humaniser* role, was the intervention to redesign the annual performance process. Among other outcomes, internal and external communication at Accenture has become much more visual, personal and story-led. Moreover, design continues to have an impact on the physical environment, making it more conducive to creative leaps and fostering innovation. All of these combined generated a strong impetus for creating change towards a more approachable, human-centred organisation.

Power Broker

'Great customer experience requires great employee experience.'
Sarah Brooks, US Department of Veterans Affairs

Summary of the role's capabilities

- Leveraging the independence of the design profession and its focus on the user as the ultimate reference point. This has the capacity to diffuse tensions and realign internal teams around a common goal.
- Utilising the power of the consumer-centric purpose to focus everyone's attention on a pragmatic solution instead of their own, fractional interests.
- Changing the frame of reference and deliberately upsetting the entrenched power structures by, for example, the implementation of new and compelling ways of working.
- Immediacy of impact of the proposed ideas, made possible through the quickly evolving prototypes everybody can relate to.
- Shifting the attention and the ultimate organisational metrics towards people-focused solutions rather than systemic or cultural problems.

The persistent focus on user needs creates a neutral reference point. It is difficult to oppose the voice of the actual user as represented by the personas or user journeys put forward by designers. Take the case of the US Department of Veterans Affairs. Highlighting the fact that the veterans often wait two months to access medical care changes how the power dynamic works at an institution. The designers' ability to bring to life users' needs and struggles serves as a levelling device, which has the capacity to bring together the different parties involved. Their vested interests are put into perspective by showing what really is at stake. Designers are the relentless advocates of the people they design for. Every single point of conflict can be brought back to the central argument that it's not about 'what we, the project team think, but it's ultimately about what is best for the people we serve'.

But why does it seem to work when the designers are the advocates rather than sales people or marketers, for example? The reason lies in the unique standing, culture and tools offered by the design profession. Other functions within an organisation tend to treat the human on the receiving end as a rational agent (e-con in the nomenclature of economics) first, a person second. They usually create a distance between themselves and the individual. Designers, on the other hand, tend to relish the closeness and radical empathy, which they use when creating products and services.

This closeness to the users gives them a much richer vantage point It also gives them a world of insights beyond what can be gleaned from questionnaires and reports. This, in turn, makes designers a more in-tune, more emphatic representatives of the ultimate beneficiaries. This is instantly apparent to other functions and professions that come in contact with them. Having established that framework of being on the same team as the users, gives the designers a unique, more neutral position versus others, who don't have that same level of human-centric legitimacy and credentials. This is what makes the design as a *Power Broker* role possible.

Another enabler of the *Power Broker* role is trust. Designers don't tend to represent themselves; they tend to represent someone else's needs. For the most part, their first priority isn't to defend their profession's tools, methods and reputation. They are more interested in defending those who they care the most about, namely the people they design for. Other professionals, because of the demands their professions place on them, have a certain reputation to uphold and are perhaps more focused on profit, process or efficiency. Designers are part of a profession predominantly concerned with the pursuit of originality, fit with user needs and aesthetic fulfilment. These are powerful and authentic motives, which, if decoded correctly by the other parties involved, create a space for trust to develop.

Tools that help to create prototypes, which designers bring to the party, have the capacity to create an immediate impact and a transparent vehicle for iterating the best possible solutions for the user. This is not what psychologists, sociologists or anthropologists can offer. They also have the tools and outlook to get close to the real people, with their intricate and multifaceted lives, but they can't make the leap into a practical, new reality. This ability to show a tangible representation of the ultimate value quickly, means the efforts to unite and mediate between the different groups are focused on the task at hand. It leaves less scope for in-fighting and divergent agendas to take hold. Bringing it all back to the central point, as depicted by the quick, working prototypes, service blueprints, personas and similar tools, helps to ensure the process is anchored in the real user needs. In case of the US Department of Veterans Affairs, the unifying tool was the veteran's journey. It focused everyone's attention on the veterans and not all the problems the organisation was facing internally. It served as a powerful reminder that the veterans were the ones really in need. This re-directed the conversation and re-tuned the organisational noise coming from the various groups.

So, when does the *Power Broker* role come into its own? There are several organisational obstacles and challenges that can be overcome by design.

A common challenge encountered by those who want to create a more customer or user-centric organisation is to be faced with resistance from the existing structures inside an organisation. Because of the nature of large institutions, internal power structures are integrated vertically and centred around functions; marketing, IT, operations, finance would have their own line of command, their own procedures and more importantly their own professional view of the world, i.e. professional culture. For example, one profession might have a very low tolerance of ambiguity and uncertainty, whilst the other might have a very high tolerance; one function might favour predictability and known industry reference points, whilst the other might favour originality. The Adur and Worthing Councils case shows a leader, Paul Brewer, who's been able to overcome the highly rigid and risk-averse culture in an effort to install a more modern, collaborative and transparent organisation. What he'd learned about service design and its impact in his previous roles, proved instrumental in shaping the successful change effort at the council. Design-inspired approach, with its focus on usability, flexibility and transparency, had a fundamental influence on overcoming existing hierarchies and creating a more effective organisation.

Another example of the *Power Broker* role in action is Telstra, where co-creation processes and workshops made sure people from different backgrounds and different professions came together. It helped to align their work and concerns around key user needs, thus diffusing fractional, internal priorities. In an engineering dominant organisation, it was a no small feat to shift the attention from internal priorities towards the needs of the consumers. Cecilia Hill and her Design Practice team were able to leverage the empathy of design to support the organisation in their aim to become more customer-centric.

Friendly Challenger

'We have to be constantly provoking, challenging and questioning people all the time–Why did you do it in that way? Is it really what you want? It this really what the client want?'
Ellen Kiss Meyerfreund, Itaú Bank

Summary of the role's capabilities

- Providing a safe haven for fledgling ideas to grow and develop without being challenged prematurely. This refers to both physical and mental spaces available within an organisation.

- Encouraging an atmosphere of openness and genuine interest in a best possible solution, regardless of its origin.

- Drawing attention away from internal politics and tensions and towards the users' needs.

- Creating an environment where it is the norm to question basic assumptions, critique each-others' work constructively and champion the search for the best possible (feasible, viable, desirable) solution.

It is not uncommon for the biggest obstacles to organisational transformation to originate from conflicts of interest and style amongst the various groups participating in the process. Confrontation, animosity, divisional and professional rivalry all contribute to the tense atmosphere that accompanies organisational change. You have probably witnessed, first hand, how progress can be hampered by an executive who is fearful of diminished standing as a result of the proposed change. In many situations the fear of losing one's position will be an inevitable part of the process. Nevertheless, change is about managing the shifting internal landscape and is always associated with some degree of tension.

Challenge is necessary, but for it to be healthy and effective it needs to happen in an environment of trust. If the motives, motivations and purpose of those who challenge the status quo are transparent and genuine, it is more likely that such questioning and prodding will be more sympathetically received. If the challenge comes from a person or a group that is suspected of having ulterior motives, the probable outcome is entrenchment and further escalation of tensions. That's why designers and design thinkers, with their determined focus on users' needs are particularly suited to the role of *Friendly Challenger*. It is their efforts to unpack, in great detail, what people actually want and desire, that firmly anchor the conversation in value creation and divert it away from internal politics and squabbles. The more the design-driven change insists on referring back to the users, the less partisanship is perceived and the more trust is injected into the process.

What helps, is the transparency fostered by the tools, methods and techniques associated with the design approach. Service blueprints, deep and rich ethnographic narratives presented in user personas, role-play to demonstrate potential interactions to detailed and emphatic customer journeys, all assist in providing a clear framework for challenging the status quo. Thanks to the extensive use of these tools it is apparent to anyone involved that the intention is not to gain power or status, bur rather to improve the offer to the end users. This has the power to diffuse arguments and direct the conversation towards pragmatic solutions.

In case of Innovation Studio Fukuoka, the role of *Friendly Challenger* was important on a number of fronts. In order to deliver innovative, socially-minded, new ventures, participants must first challenge pre-existing assumptions. During the ideation stage, the ability to offer honest criticism to each other offered a valuable tool within a group dynamic setting. This point is particularly important since the Japanese are generally reticent in offering direct criticism. The atmosphere generated during the process was characterised by generosity and warmth, which enabled for the critical comments of the mentors to be received in the spirit of learning and idea improvement. And finally, the project team often ended up becoming critical friends to some of the participants, offering them additional support after the project ended.

At Itaú Bank, design has seen its remit extend to questioning existing practices and acting as an agent provocateur. Through specific initiatives, the Innovation Team has set out to challenge received banking wisdom and push the envelope of what's possible in this highly regulated sector. It is a radically different approach to what's normally found in this part of the economy. One such initiative was the Challenge competition where participants, i.e. the employees, were encouraged to let their corporate imagination run wild. It provided a necessary 'safe space' to innovate and to propose riskier ideas to the business. It insulated those taking part in it from the usual, harsh scrutiny of the highly structured industry and allowed them to propose more experimental solutions. The creation of special project spaces called Inovateca could also be seen through the lens of the *Friendly Challenger* role. In this instance, dedicated spaces, which could be personalised by project teams, acted as safe heaven to push the envelope of existing ideas.

Technology Enabler

'The way that we communicate about ourselves is completely different from the way that normal space companies communicate and present themselves...it's the design team who really captured who we are and has made our voice much more natural and meaningful to sectors outside of the space community that we want to be engaging with.'
Stuart Martin, Satellite Applications Catapult

Summary of the role's capabilities

- Making technology useful by emphasising the usability of the systems in place in order to maximise engagement, reduce errors and increase satisfaction of the systems' users.
- Ensuring smooth workflow between physical and digital platforms.
- Making sure employees' needs and expectations are catered to, not simply the technical system requirements.
- Focusing on the usability as well as the aesthetics. Since people come in contact with certain technologies for extended periods of time, the aesthetics of these technologies play a key role in productivity as well as overall job satisfaction.
- Supporting the buy-in, adoption and continued usage of the system–people are put in a situation where they want to use the technology provided to them not because they have to, but because they want to.

In contrast to the popular view that says the more technology the better, too much of the wrong kind of technology can often be disruptive and overwhelming. Clunky, complex, opaque, ubiquitous, demanding–these are just some of the less appealing characteristics of many technological systems that people in organisations come in contact with. Good design, which centres on human needs, ergonomics and interaction principles, has the capacity to make technology really useful, pleasant and even fun to work with.

Where design comes into its own is in making technology accessible and usable. It's quite extraordinary that in the age of the iPhone so much innovation is still technology-led rather than human-focused. This classically stems from too much focus on engineering capability being put in charge of the design of the product. This almost always leads to badly-thought-out solutions that are generally not fit for purpose. The situation is particularly acute in engineering-led, B2B industries. With that in mind the case of Satellite Applications Catapult is quite noteworthy. As a space industry company they've decided to fully embrace design as pivotal agent to make their technology solutions and thinking accessible and versatile.

Since the introduction of Scrum/Agile as the leading software development methodology, the focus on the user throughout the process has been much more prominent than in the past. It is now common practice to gain users' feedback early and often, thus ensure that ultimately the final setup will add value for the users. In our experience, in large part due to the ubiquity of the Scrum method and the Agile way of working, IT professionals are getting better at empathising with the users and delivering what they need. They are not, however, trained to the same depth as designers when it comes to gauging people's minute desires and subtle preferences. The latter group, when asked to provide a solution, would relentlessly focus on the people and be less concerned about technology as the starting point.

Digital transformation is often used as a 'soft' rehearsal for a larger more ambitious organisational transformation. This was the case at Adur and Worthing Councils. They needed to overhaul their IT system, which was expensive to maintain. They took the opportunity to use the digital transformation project to change the way employees worked and embed a more human-centred design into how they think about, design and deliver services. In order to encourage more collaboration and transparency in how they worked, they moved to the Google for Work service and put together an inexpensive eco-system of existing solutions. An IT-led initiative might have placed more emphasis on building a bespoke system or adopting a costly, off-the shelf package. What actually happened was a shake-up of the received working paradigm–from opaque and silo-centric to transparent and collaborative.

We believe that the trend of enterprise software following the consumer software when it comes to usability, functionality and aesthetics is irreversible. People in corporations and businesses expect the same level of polish and ease of use from the platforms they use at work as they do from those they use at home. The popularity of apps such as Slack or Yammer are testament to that. Designers and design thinkers understand this and are driven by how to make technology connect with the human being at the other end. This is what makes them invaluable when it comes to organisational change, which currently almost always requires an element of technology.

People, except for the digital native generation perhaps, are naturally reluctant to constantly learn and re-learn new technologies that are coming online at an ever increasing pace. The fear of an unfamiliar user interface and user experiences is real. It is sometimes mistaken for the fear of technology itself. This is where the design approach to change comes in. It ensures that people are eased into the new interactions as fluidly as possible. This, subsequently, reduces employees' fear and increases the chances of a successful transformation.

Community Builder

'The programme involves specialists and advisors from industry, academia and mentors, as well as attracting people from all walks of life. As a result, the growing network of participants developed through the programme has a really big impact on their lives.'
Yuki Uchida, Innovation Studio Fukuoka

Summary of the role's capabilities

- Creating conditions for the community to come together by providing a safe, open atmosphere with the users and people involved.
- Providing tools and techniques that offer an instant feedback loop for the participants to respond to, in effect creating a fuel for the community to coalesce and work together.
- Ensuring a level of empathy, which enables the designers to connect with different constituents in the community on a deep and intimate level.

The need to foster cohesive, thriving communities has never been greater. Societal, environmental and economic pressures mean that solutions must be coordinated, agreed and acted upon collectively. If we are talking about a new school, an innovation centre, a large governmental investment, or any other initiative that requires bringing together disparate needs of multiple parties

and stakeholders, design-driven engagement offers a good model to follow. It creates a safe and inclusive environment where everybody is welcome. It draws on the strengths of the other roles, including *Power Broker*, *Cultural Catalyst* and *Friendly Challenger* to provide a platform where different groups can build on each other's ideas and where the plurality of voices is encouraged and appreciated.

Design-driven approach is usually very hands-on and involves training sessions, workshops and live prototyping, where members of the community can create their own models and scenarios. This process-user intimacy creates a level of engagement, which is different from a typical community consultation. Through co-creation and other forms of collaboration, a genuine interest and curiosity are ignited and a natural willingness to participate emerges. As the empathy displayed by the designers creating the space for dialogue is apparent to everybody involved, and the fact the efforts usually centre on the needs of the community members themselves, the efficacy of the process is greater than it otherwise would have been.

In the case of Innovation Studio Fukuoka, providing focus on the user and thanks to the apparent professional agnosticism, designers were able to create a platform for the disparate organisations and agendas to come together. Entrepreneurs, large organisations, SMEs, academics, students, government and charities rarely have a chance to interact and work towards a common social goal in this part of Japan. The platform provided them with a means to do just that. A traditionally reluctant community was effectively encouraged to participate and contribute with ideas and solutions. The result was vibrant participation and an oversubscribed list of entries as well as several new business start-ups.

The case of Itaú Bank highlights how a design-driven approach helped its people to think and act differently. Mobilising people to learn by doing was one of the ways the organisation was being transformed. A short, 5-day sprint served as a means of putting people in the right mindset and challenging them to come up with something new and different. Spotify has a similar initiative called Hack Week where everyone is given a week off to work on something they are passionate about. Organisational structure is also important to ensure design supports the building of an innovative community. Spotify did this by re-balancing the design capabilities with the tech and product functions and they now have a more balanced community to grow from.

Community building does not just involve bringing people together, you also need to offer them a space to come together and collaborate. Fjord was fiercely protective of their studio spaces and insisted that Accenture protect and invest in project spaces for this very reason. Satellite Applications Catapult's 'Design Cave' and Itaú Bank's Inovateca spaces are similar types of protected creative spaces.

Section 1
Case Studies

Steelcase

Accenture & Fjord

ZOOM Education for Life

Deloitte Australia

Bumrungrad Hospital

Spotify

Itaú Bank

Telstra

US Department
of Veterans Affairs

SAP

Adur and Worthing Councils

Satellite Applications
Catapult

Innovation Studio Fukuoka

Steelcase: Reimagining the future of work and workspaces

Introduction

Steelcase is the world's largest maker of office furniture based in the US. It celebrated its 100th anniversary in 2012 and is one of the longest running organisations featured in our book. It is also the most mature in terms of using design as an innovation and organisational change tool. Design has been a key market differentiator for Steelcase. Although competitors like Herman Miller may be better known for classic, modernist design; Steelcase has really used design and specifically design thinking to position itself as an insight-led company that focuses on addressing future needs of the work place.

Steelcase has in recent years started to explicitly leverage their design-led culture and expertise in a number of ways. They began designing "WorkLife" centres around North American and later globally, in 1995. These locations focused on creating "experiences" for clients and staff that brought the changing nature of work to the fore. They have established innovation centres, the first was at their global headquarters in Grand Rapids, Michigan in 2013 and more recently another centre was established in Munich, Germany in 2016. They have been using design to drive an insight-led strategy internally as well as offering a human-centred design approach as part of their consulting services to external clients. Although Steelcase's relationship with design has been well documented over the years, this is a good time to revisit their story and take stock of where they go next.

Who we spoke to
Izabel Barros, Head of the Applied Research & Consulting team in Steelcase Latin America
Donna Flynn, Vice President, WorkSpace Futures
Dave Lathrop, Director, Applied Research Network

Why change?

The story of Steelcase isn't so much about 'why change', but about how they continue to rejuvenate and be at the cutting edge of the workspace sector. Steelcase has always been an innovative firm, starting with their first metal wastebasket launched in 1914 to when they initially started using design as an innovation tool, to adopting design thinking as core practice throughout the organisation. As a result, Steelcase has considerable experience using design (in its various roles but more notably as a *Cultural Catalyst, Humaniser, Framework Maker, Power Broker and Community Builder*) compared to other examples in the book. The Steelcase example illustrates that transformations doesn't just occur once or twice in the lifespan of the organisation, but are a constant feature in any healthy and thriving organisation. Design has not only become a market differentiator for Steelcase, it has influenced they way they operate and most importantly how they manage continuous change in the organisation.

Design roles that enabled change in Steelcase

Types of changes achieved through design

Since 1997

Changing products & services

Changing organisation

Changing the process of change

What has a design-driven approach brought to Steelcase?

- Learning and embedding design thinking in their business language gave Steelcase a competitive edge.
- Design helped formalise and cement Steelcase's historical focus on being human-centred. It provided them with a framework to turn user insights into market leading strategies, services and products.
- Steelcase is using their knowledge of design thinking and offering it as part of their service to clients.

Roles of design in the last 20 years

Steelcase is an interesting case study as it allows us to learn how an organisation continually practices and leverages their design capability in an organisational context. Steelcase is also a powerful example of how an organisation activates design at various points of their journey and illustrates the long-term appeal of design's role as an innovation tool and more importantly as an organisational design tool.

When Steelcase was first introduced to larger ideas of design through IDEO in 1996, they were very much using design as a *Cultural Catalyst* to help the company evolve from it's manufacturing roots to one that is more idea based. In the early days when design was being introduced, the design process functioned as a *Framework Maker* to enable innovation practices to be set up and embedded in the company. The questioning aspect of design (activated in a *Friendly Challenger* role) can be seen in Steelcase's central practice in requiring that every project must have a central question.  And since its adoption of design thinking, the underlying role for design at Steelcase has been to humanise what work means for individuals. Design has also been used by Steelcase as a *Community Builder* to help build communities of practice through collaboration and co-creative approaches. This is especially evident in the way they work with internal teams as well as with external clients.

Steelcase's use of design in the last 20 years has moved from a product development function to one that is used to influence the organisation itself as well as influencing the very process of change by helping the organisation overcome obstacles and resistance to change. They have been successfully leveraging their experience using design thinking internally to offer design as a service to help clients develop innovative workspace solutions.

Design thinking is the core mechanism by which Steelcase considers how it does what it does and we look to the work done by several influential groups in Steelcase: Product Development, WorkSpace Futures and Applied Research and Consulting to illustrate this key aspect. Through their work, we also show how design can continually transform and help an organisation keep at the forefront of change.

The changing context of workspace

The way we work and our perception of work has certainly undergone a significant change in the past 30 years. Part of the reason includes the changing perceptions that work is life and work isn't separate from life. Work-life integration has replaced work-life balance since a big part of our life is work and we also want that part of our life to be an amazing learning experience that's healthy and rejuvenating for us. There is a growing recognition that how

Stages of transformation

1. The introduction of design thinking through IDEO in 1996.

2. WorkSpace Futures group evolved from traditional R&D function in 2002 to focus on using user research to gain insights into future work trends and define a future vision of workplace environments. In parallel, the Applied Research and Consulting group was set up in 2001 to offer design research insights and user centred design expertise to clients.

3. Donna Flynn took over the WorkSpace Futures group in September 2011 and expanded their influence across the product cycle as well as working on longer term R&D research that seeks to understand the future of work and the impact it has on human experiences of work.

What can we learn from Steelcase's story?

Change is the only constant. Transformation occurs at various points and for various reasons. Being a design-led company has enabled Steelcase to anticipate challenges and address them successfully.

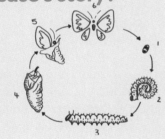

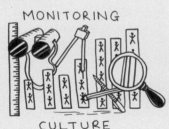

Organisational culture always shifts and evolves, requiring constant awareness, support and encouragement.

When an organisation is design-led, design as a concept gradually becomes less discussed. The need to label processes and activity as 'design' falls away when it becomes inherent in everyday practices. However, this can also become a challenge when design becomes too invisible and there is a need to 'reintroduce' design if it becomes too diluted.

we work and where we work is a really important part of our lives.

At the same time technology has also been massively influential in changing the work experience. The ability to do progressively more aspects of our work, from anywhere at any time with anyone, has challenged our perception and expectation of what our working environment should be. The workplace is becoming more distributed, connected and collaborative. At the same time, there is also the need to help people switch off and find private spaces to reflect, concentrate and be creative. These sometimes conflicting requirements are current challenges that Steelcase seeks to understand and anticipate in their work.

Insight-led design culture

Having the user at the centre has always been part of the Steelcase story and this focus on the user has helped transform it from traditional manufacturer to industry innovator, recognised as much for their insights in the work environment as for the products themselves. Its perhaps unsurprising that Steelcase should be instinctively drawn to design, which naturally embraces deep empathy, and a quest for the most profound insights linked to the intricate and often hidden needs of the users.

Steelcase's first product was born out of user observation that people often would throw paper as well as cigars into wastebaskets. This insight led to the creation of a strong, durable but low-cost fireproof wastebasket that significantly reduced the threat of fire. It was considered a major innovation way back in 1914. It's this focus on users and a deep understanding of how behaviours influence the way we work that has kept Steelcase at the forefront of its market and probably the key reason why it took to design so well.

The focus on user behaviours and designing according to user needs has always been part of Steelcase's DNA. It's therefore unsurprising that the company has a history of patents. Ideas are welcomed and everyone in the organisation is encouraged to participate in the innovation process. Saying that, the link between user research and innovation wasn't always clear and design wasn't always at the forefront of the business.

That changed when Jim Hackett became Steelcase's CEO in 1994. When Jim took over at Steelcase, the company was struggling after years of falling revenue and they reported a $70-million loss for that year. It was a challenging time for Steelcase, the furniture industry was suffering a downturn and this was followed by a national recession. Historically it was a time when manufacturing was in decline and Steelcase were shutting factories down, shifting manufacturing overseas and moving to a more globally integrated model. It was in this context that Jim discovered design thinking through his friendship with David Kelley and Bill Moggridge (founders of the famous international design and consulting firm IDEO). He saw the value in design and realised that the future of Steelcase was not just in products, it was in ideas and this became the major driving force of Steelcase when Jim took over.

Jim was also inspired by the way IDEO worked, which was non-hierarchical and team-based. Being exposed to a different way of working inspired Jim to change how he viewed the industry and Steelcase's potential role. He started to move the focus of the business from selling work stations based on a dominant way of working to one that is future focused and team-oriented.

The extent that Jim Hackett believed in design and how it can help them transform led Steelcase to became the majority stakeholder in IDEO. The firm's CEO, David Kelley, became Steelcase's vice president of technical discovery and innovation. Through the influence of Kelley and IDEO's culture, design thinking started to permeate Steelcase's culture. Kelley's involvement helped establish a more sophisticated understanding of design in the company and cemented its importance in Steelcase. Although Steelcase's involvement with IDEO diminished after 2007 when Steelcase sold the majority of their shares back to IDEO's managers, they have maintained a working relationship over the years. Steelcase also has rich global research partners ranging from academic (e.g. MIT and Kyushu University), corporate (e.g. Xerox PARC and Arup) to independent organisations (Fraunhofer Institute and Santa Fe Institute), to encourage continuous cross-pollination and diversity of thought in the company. These reasons as well as the widespread executive adoption of design thinking and being human-centred and insight-led to this day has meant that Steelcase continue to grow a design-framed culture, which remains extremely rare and unique amongst corporations.

What makes a company design-led?

While many of the organisations featured in our book are attempting to create a more human and innovative culture through design, Steelcase is a great example on how to use design to achieve this. We asked people we interviewed what constitutes a design-led culture. For Donna Flynn (Head of the WorkSpace Futures group), it is the democratisation of design and they way its embraced even by people without a design background.

One of the key differences in working in an environment that has a more mature understanding of design is that it takes less effort to convince people of the value of design research. For Donna, this was a key difference between where she worked before, which was a very engineering-centric culture, and a culture that really values design thinking.

'One of the interesting things when I started at Steelcase was the difference in attitude towards design research. Where I was before, I had to keep banging on people's doors to get them to understand the value of design research. But when I came to Steelcase, everyone got it straightaway. So for me personally it was like heaven because I thought, I don't have to spend all my time explaining what I do, we can just do it and have more impact and gain more influence with the business.'
Donna Flynn, Head of the WorkSpace Futures group

For Dave Lathrop (Head of Applied Research Network), the single biggest signifier of a design-led company is how it makes decisions. According to him, organisations are generally either structurally led (process driven) or they are culturally led. A structurally led company needs to 'design' itself with an eye toward inculcating the behaviours it desires. A culturally led organisation has people who can "know" what to do because they are immersed in a design thinking system. And design for him is an incredibly powerful approach to empower people to know what to do and through this, to subsequently unlock the human potential of people within an organisation. In this instance, Dave is very much referring to design as a *Framework Maker*, offering not only a common language and process but a mindset focused on people and the future.

Using design to influence the organisation and the process of change

The way design thinking is used and applied in Steelcase manifests in a number of ways and through different teams in Steelcase. Although design has always been a key feature in their innovation team and product development processes, we are particularly interested in understanding how design is used to influence the organisation itself and how it influences the process of change. For this reason we have chosen to focus on two groups in Steelcase, WorkSpace Futures and Applied Research and Consulting.

WorkSpace Futures Group

The WorkSpace Futures group evolved from R&D in 2002 and was initially led by Joyce Bromberg to study workplace trends. Joyce was instrumental in refocusing user research to gain insights into the future of work and to define work trends and a future vision of workplace environments.

The group in its current incarnation has changed and developed over the years. When it was first set up, Joyce was responsible for user-centred research for vertical markets and early phase research for new product development. Donna Flynn (an anthropologist by training) became the director of the group in 2011 and has focused on developing the group into a world-class research team. The team is now focusing on complex and challenging problems that will change the experiences of work, worker, and workplace in near and distant futures. This is important in order to develop insights that point to possible and probable futures that can impact Steelcase's strategy and the needs of its customers. As a result, they cover a range of topics, from advances in the neuroscience of attention, learning, or creativity to understanding how workplaces can be designed to support human wellbeing and thus positively impact employee engagement. They also look at work practices across a variety of environments: corporate workplaces, small businesses, healthcare and education.

They have distributed global teams located in Hong Kong and Europe, with the core team based in Grand Rapids, and a few team members in Colorado and the Silicon Valley. The WorkSpace Futures group has around twenty-two members comprising social scientists, design researchers and technologists.

Cultural (Re)Catalyst

The WorkSpace Futures Group is quite unique in the way it operates within Steelcase. They have the freedom to set their own design research agenda outside the constraints of the core business. This allows them to consider a longer time span and anticipate future scenarios. Every three to five years they refresh a set of research themes they are exploring – such as 'Humans + Machines: How might we anticipate future combinations of humans + machines and design systems to elevate uniquely human promise at work?' and 'Embedded

Learning: how might we situate learning as a continuous practice to accelerate change for individuals and groups in complex adaptive systems?' Knowledge is developed around these themes through a mix of different types of projects, including secondary and primary research, behavioural prototyping, or projects with external research partners.

This approach enables the business to think longer term without the day-to-day pressures of product development timelines. Standard business pressures might limit the scope for proper exploration on projects, hence reducing the actual reach and meaning of the design-led values. For Steelcase, creating a space and framework that enables the more exploratory nature of the design thinking to be fulfilled, meant that the core of the design-led culture was revitalised and carried forward. Here design acted as the Cultural (re)Catalyst since in Steelcase's story it's not about catalysing a new culture but offering a 'safe' space for the fragile but crucially important aspects of a design-led culture to grow.

A framework for innovation

In the last three years, the WorkSpace Futures group have expanded their influence across the entire product cycle. Although the work produced by the group has been really successful and well received, it was still predominantly user-centred research that was being applied at the back end of the product cycle to inform marketing messages and sales. Donna's task was to diversify the research in order to have more influence at the beginning of the product development cycles and to provide more strategic impact on the whole business. She shifted the existing expertise and capabilities to bring more value across the business, including front-end innovation activities, business strategy development, and even corporate functions such as Human Resources or IT.

The WorkSpace Futures group partners very closely with the leaders of the businesses and with the design and marketing teams that are supporting the businesses. These two modes of engagements require a careful balance in priority setting, not only do they have to consider current trends but also to anticipate longer five to ten year trends. They are continuously filtering their priorities through these two lenses and this results in having some of their researchers being embedded in different parts of the business that are more specialised for example in healthcare and education, while some researchers are based regionally in order to understand the cultural aspects of work that might be different in various geographic regions.

The design thinking culture at Steelcase advocates that every team leader develop an insight-led strategy. For this reason they partner closely with the WorkSpace Futures group to help them prioritise, develop, and frame the core insights for various business and product strategies. Connected to this, the group is also involved in strategy, operations, human resources, etc. often operating as something between a resource and a coach. In addition to this they work very closely with brand communications to develop both corporate communications stories as well as product messaging. All of this helps to deliver a strong, insight-led customer experience, and one of Steelcase's key differentiators with regard to their competitors is the depth of insights behind the business.

What are the conditions for impact in Steelcase?

- An organisational culture that affords a focus on people.
- Long term and consistent support from the executive team of the new behaviours that gradually become the culture within.
- Have teams dedicated to using and spreading design methods.
- Executive sponsorship of design thinking as a practice.

What have been the challenges so far?

- Creating and sustaining a design-led culture.

What type of change still needs to be achieved?

An on-going challenge is the learning of design thinking for new employees—especially for more senior people as they transition into the culture. It's learning a new language and way of interacting that is a key part of the culture.

Applied Research and Consulting Group

While most of WorkSpace Futures' activities are focused on internal teams at Steelcase, the Applied Research and Consulting (ARC) group was set up to explicitly work with external companies. In a sense, ARC is similar to SAP's Design and Co-Innovation Centre where a design-led organisation has leveraged their design expertise and insights and offered it as an external service. The key difference in their approach is that they use their own domain expertise (place making for ARC and software development for SAP) to help organisations change.

ARC started life as the Workplace Strategies group in 1998. At that time Steelcase already had a well-established research culture and process and they decided to offer their expertise to external clients in a consulting capacity. ARC now incorporates design thinking methodologies and approach to provide consulting on physical and behavioural change in the workplace in order to help companies create a culture of innovation.

ARC has a globally distributed team of 30 people. They have offices in Latin America, Asia and Europe. Their modus operandi is to use local expertise to support senior experts in order to understand the specific cultural aspects of the workplace. ARC's mission is to optimise behavioural and cultural performance by taking a strategic view of the workplace. ARC's focus in every project is to understand how both place and place making processes can be vehicles for intentional change.

ARC utilises a holistic human-centred approach that is designed to impact the systems at work within an organisation and at the same time engages people at all levels of the organisation. When we interviewed Izabel Barros, ARC Lead for Latin America, and former member of the WorkSpace Futures group, she spoke about how they help people understand their needs, and how to better support them, and the organisational purpose.

ARC transformational approach has four distinct stages. Starting with 'Plan', co-construction sessions are conducted to review scope and fine-tune strategies to corporate culture. The general outcomes usually involve setting tactics, timings and responsibilities. Moving on to 'Diagnose and Define', they first set out to determine the context, intent and success metrics using leadership workshops and interviews. Then, 'Engage and Develop' involves ARC working to engage employees at different levels through workshops, surveys and photo missions. They observe and study people at work in order to yield insights and action steps. They foresee what the future could be through co-design. Finally, there is the 'Deliver and Measure' stage where ARC offers direction and support with coaching, training and protocols. Results are measured qualitatively and quantitatively, and adjusted to requirements.

ARC uses a number of approaches and tools depending on the needs of the project, including ethnographic approaches, surveys, direct observation during field work, workshops, making sessions, as well as scripted and open design interviews. Perhaps most potent are the participatory exercises where users explore ideas in an informational and physical way, pushing abstract ideas and changes into tangible prototypes. These classic design thinking approaches provide tools and techniques that give possibilities a clearer expression for further dialogue, refinement and alignment. This fuels the community to coalesce and work together, using design in a *Community Builder* role.

Using design to broker relationships and build communities

It's clear that ARC uses design as a *Power Broker* and *Community Builder* by bringing people from different parts of the organisation to work on a problem together. This happened for Boeing, one of the largest plane manufacturers in the world. An earthquake in Renton Washington in 2000 rendered many of their traditional engineering office buildings unusable. While this was devastating for the company it was an opportunity for Carolyn Corvi, the General Manager of the 737 platform to introduce a new way of working. The traditional way of separating the engineers and office workers from the mechanics and manufacturing employees had resulted in a divided culture with a huge gap between the people who design the product and the people who make it on the factory floor.

Corvi saw the opportunity to get people working together in a better way. Steelcase and the architectural design firm, NBBJ were asked to find ways to solve this problem. ARC assembled a team of 35 Boeing executives, managers and engineers for a three-day workshop at the start of the project. They asked participants to photo document their offices, looking for things that did not support their needs and work patterns. This visual, ethnographic approach coupled with interviews, workshops and first-hand anthropological techniques allowed ARC to tap into the knowledge and creativity of Boeing's people to reveal issues often hidden in the day-to-day activities. This process was the starting point in bringing different people together, acting as a *Power Broker* and *Community Builder*. The process of user engagement extended into many workshops bringing disparate groups together to explore new work approaches, video and in situ field observation by participants, extended piloting of new ideas for space and technologies, etc. Finally it resulted in a solution that moved 2500 employees (engineers, mechanics, executives) into a newly completed workspace towers inside the factory. The focus was on the plane, people and processes of building them. In conjunction with other production system change, Boeing now achieves 50% more efficiency (it now takes 12 instead of 24 days to complete a 737). The culture in the plant has completely changed. The engineers work in the mezzanine-level directly in sight of the planes being built and are on the production floor regularly. As a result relationships are stronger and understanding is deeper. Problem solving is real time and quick. Physically, the production and engineering staff now share areas, like mini-libraries and break areas, which help teams mingle and create a sense of community. What started out as a straightforward facilities change resulted not only in increased efficiency through earlier detection and resolving of problems but importantly a greater sense of community and shared values.

Another example of how ARC is using design to help influence the process of change is evident in the project with Graña y Montero, a publically listed engineering and construction company in Peru. What started out as a family business had grown significantly in the last few years, with 26 subsidiaries structured in four business areas. ARC helped them think more strategically and globally about their business as they expanded and became publically listed on the New York Stock Exchange. ARC helped them create a model that was able to accommodate that growth without any significant changes. It was focused on building resilience for the physical solution as well as building flexibility to enable the growth. In this instance, design served as a *Framework Maker* creating the scaffolding necessary for bringing the vision of the company to life.

Growing a knowledge network

Steelcase has been at the forefront of using design thinking and design research to understand workplace trends and to help organisations change. Recognising this, Steelcase is keen to pull together what they have learnt in the past 20 years. The newly formed Applied Research Network is an attempt at doing this and is led by Dave Lathrop. Dave was a founder of ARC and in addition has spent 20 years in WSF. As ARC is a distributed group working on projects across different geographies, one of Dave's key roles is to bring together learning and experiences into the group's shared mind. For example, project records were somewhat scattered, while practice and client learning had not been coalesced. Making the extensive learning from hundreds of past projects more explicit was an obvious first step. This leads naturally to a variety of knowledge domains that, in Dave's words, can grow to inhabit a 'knowledge garden' that not only involves internal people but is also built collaboratively with Steelcase's clients. This is yet another example of using design as a *Framework Maker* and *Community Builder* whereby the project at hand is held together by a well thought-through, human-centred system.

Related to the idea of a knowledge garden and working collaboratively to build knowledge, Steelcase is increasingly looking towards developing an eco-system involving key partners as their next step of development. One of the key culture changes in Steelcase is evident in their growing embrace of a more open innovation model. Previously, innovation happened behind closed doors, and it was generally not shared broadly. However with technology and social trends changing so rapidly, Steelcase (like many other organisations) are finding that it's far more valuable to share the building blocks with key partners than to work independently. Moving towards a more open innovation model has required them to rethink what IP means and at what point is it shared in order to gain value and learn from their innovation partners. They are looking at ways to leverage this eco-system and to push innovation at a more rapid pace.

Maintaining and renewing a design-led culture

Steelcase's story has shown how, by maintaining and renewing the momentum at the heart of a design-led culture, change can be driven from within as well as from the outside at the same time. By developing a culture that is driven by respect and empathy for each other, it automatically extends to their interaction with the outside world and to clients. Design's longevity in Steelcase is testament to its value to organisations. Design has not only become a market differentiator for Steelcase, it has influenced the way they operate and most importantly how they manage continuous change in the organisation.

Developing a strong culture in an organisation is extremely difficult as is often demonstrated through our case studies in the book. However, what is even more challenging is maintaining and nurturing that culture, particularly in times of great change and in new conditions that challenge assumptions. There are always potential dangers for the culture to 'slip back' for a number of reasons. Often the rapid growth of an organisation (see for example the Spotify case study) could very quickly change the dynamics and intimacy of the team. Sometimes it can be the pressure to meet external deadlines that leads to more focus on short-term goals (hence the idea to ring-fence the more open-ended quests in the WorkSpace Futures group). We also see examples where even the best run organisations succumb to the power of the codified structure and processes in an attempt to manage innovation practices.

By persistent attention to the behaviours that become culture over time, Steelcase has tried to rejuvenate its culture through the various efforts described in this case study. Above all, it's important to encourage an on-going dialogue so people understand the direction of travel for the organisation and are able to interpret them for themselves. For Steelcase, they continually strive to uphold a design-oriented social system by being curious, connected and committed.

Accenture & Fjord: Placing design at the heart of organisations

Expanding Accenture's design capabilities

Fjord and Accenture is an example of how design has been used as a *Cultural Catalyst, Humaniser, Friendly Challenger* and *Community Builder* to help transform a global consulting service into a 21st-century, professional services business equipped to help their clients and themselves deal with continually disruptive markets. Design has become the missing piece in the corporate world, bringing humanity into business strategy and technology. This case study also helps us to understand what the challenges are in embedding a design culture in an organisation from the outside. Other case studies in the book offers examples of how design can be embedded from the inside whereas Fjord and Accenture's example illustrates what happens when two different cultures merge.

Accenture is a leading global professional services company providing strategy and consulting services in digital, technology and operations. They acquired Fjord, a London-based global design and innovation consultancy in 2013. The surprise acquisition expanded Accenture's design capabilities through its existing Accenture Interactive group. The acquisition was aimed at complementing the marketing, content and commerce services offered through Accenture Interactive.

The acquisition is part of a current trend that sees consulting service companies, like Accenture and Deloitte Australia, acquire and grow design capabilities in order to help their clients through changing market conditions.

Who we spoke to
Olof Schybergson, CEO, Fjord
Abbie Walsh, Managing Director, Fjord UK and Ireland
Shelly Swanback, COO, Accenture Digital

Why change?

Consulting and tech services firm Accenture has a lot of experience in organisational change, operating model design, strategy, HR and IT transformation. They have strong business and technical functions but  needed design to drive sustainable change for their clients through a more human-centred approach. They have been steadily building up their design capabilities through Accenture Digital but realised they needed to grow their capabilities quickly and more directly through the acquisition of Fjord, a design and innovation firm.

Design roles that enabled change in Accenture & Fjord

Types of changes achieved through design

Since 2013

Changing products & services

Changing organisation

What has a design-driven approach brought to Accenture & Fjord?

- Changed the way they work with clients, especially regarding innovation and managing change.

- Helped them become a more human organisation.

- Redefined their performance management system and process.

From a rebel to part of the establishment

Fjord is a design and innovation consultancy founded in London and has been pushing the boundaries of design since they were founded in 2001. Even after the dot.com bubble burst in the early 2000s, the founders of Fjord: Mike Beeston, Olof Schybergson and Mark Curtis, were committed to focusing on the digital medium. They were also the first few design consultancies to move sideways from a communication and web design portfolio to designing services and solutions for people, leading them to service design. They had also been working on mobile platforms since 2001, five years before smart phones were released in the market. This early experience with mobile and their focus on design, innovation and digital gave them the foundation to meet the growing demand when the market turned in 2008. So you could say that Fjord has always been ahead of the curve in terms of where design practices were heading.

But the focus of Fjord hasn't changed throughout its growth. They have always been interested in the intersection between design, digital and where innovation meets mass market appeal. They consider themselves to be very much a design-centric organisation and their mission remains 'Design at the heart'. As Olof Schybergson, the CEO and co-founder of Fjord explains:

'Design at the heart means to aim at the heart of the user. You make sure that you design something that isn't just a utilitarian solution but actually really meets their expectations and their emotional needs so that you have a better opportunity to develop a lasting relationship. We also like the idea of putting design at the heart of society, making a meaningful impact and making design the heart of our clients' organisations. It is also relevant to our current context, putting design at the heart of Accenture.'

Putting design at the heart of Accenture

Lets explore why Accenture wants design at its heart. Accenture's decision to acquire Fjord in 2013 was a surprise but not unexpected considering the recent trends where global businesses in different sectors such as Capital One (financial) and PWC (accounting) look to expand their capabilities into design and digital. Large organisations are beginning to realise that in order to enact transformative change in organisations, and to a larger extent in society,

it's important to have three perspectives: technology, business and people. This refers to IDEO's model of Feasibility (technology), Viability (business) and Desirability (people). Increasingly design is becoming the leading factor in this model.

Stages of transformation

1. Accenture Interactive was already in existence as part of the capabilities group in Accenture, alongside Accenture Mobility and Accenture Analytics. Accenture Interactive offers end-to-end capabilities—including creative design, customer experience and ecommerce development—that deliver enduring customer relevance at scale.

2. In 2013 Accenture acquired Fjord as part of their plan to grow design capabilities in Accenture. The acquisition plan also included Acquity and avVenta to enhance the capabilities of Accenture Interactive across digital content and eCommerce platforms.

3. In 2014 Accenture Digital was created combining digital marketing, analytics and mobility.

4. In 2015 Accenture revamps its Annual Performance Monitoring system in favour for a more reflective, personal and responsive approach. Fjord was instrumental in helping them design the process and create new tools to support it.

 -> >  -> >

What can we learn from Accenture & Fjord's story?

A design-led culture doesn't necessarily need to emerge from within the organisation.

RESPECT BOUNDARIES

Using design to drive organisational change requires resources, capabilities and reach not usually present in a traditional design company.

There has to be mutual respect and boundary setting when trying to merge cultures.

Accenture's interest in design came from two main reasons. First, they had been collaborating with Fjord in previous projects and had experienced first-hand the impact it made. Second, they were seeing demand for design and design thinking capability through their clients. As a very client-focused organisation they were keen to ensure they addressed this gap by putting design at the heart of the organisation.

So there was a clear synergy between Accenture and Fjord's vision. For Accenture, it was a way to bring in the humanising and creative perspective so important to organisations. Design in this context was used as a *Humaniser* as well as a *Cultural Catalyst*. Accenture's aim was to use Fjord (its people, practices, cultures etc.) to catalyse a more creative and human focus in the organisation. For Fjord it provided them with access to cutting edge technology, data analytics and business and industries capabilities. For Accenture, it offered the opportunity to leverage these three perspectives (technology, business and people) at quality and scale.

Merging pains

Merging two very different cultures was always going to be challenging. Although there were obvious synergies in the mission and values of the two organisations, there were obvious differences between the corporate culture of Accenture and the creative culture of Fjord. The first year of the merger was especially challenging. At a fundamental level, helping Accenture understand design and the role it can play was the first task. Since design

has never been a leading function in Accenture, it was important to work out what practices, habits, processes and methods were similar and how they were different. For an organisation with more than 360,000 employees, it's a huge challenge to build this understanding and there is a continual need for education, collaboration and encouraging people to remain open-minded about different perspectives, starting points and approaches. To put it into perspective, Fjord had 200 employees when it joined Accenture, which is only 0.05 per cent of Accenture's workforce. This imbalance in scale was initially quite daunting for Olof and other Fjordians. There was a lot of soul searching in the early days to work out Fjord's role and identity in this global organisation.

'Initially we were naturally a bit daunted and did not know what to expect when the merger was announced. Accenture is so big and different from Fjord. We were acquired into Accenture Interactive, which is a small sub-set of Accenture, and actually probably the most culturally similar. They had a similar mindset and that helped us settle in. But ultimately

*we're fundamentally stronger and bigger and better and doing much more
impactful design work now than we've ever done.'*
Abbie Walsh, Managing Director, Fjord UK and Ireland.

Protecting and enabling a design-led culture

Fjord took pains to ensure that their culture,
carefully cultivated since their inception
in 2001, was protected when they joined
Accenture. It was important to agree to what
they needed to protect. Brian Whipple, who
leads Accenture Interactive, asked Olof to
name what the 'non-negotiables' were when
they first discussed the possible merger. These
non-negotiables were elements of Fjord's
culture that should not be changed. For
example, one of the non-negotiables agreed

early on was to maintain and invest in their physical studios. Olof knew having
the right type of environment (physical as well as culturally) was important to
bring a diverse project team together, which would also include clients and
their Accenture partners, to co-create and collaborate. What made the merger
work was the continuing, open dialogue about the non-negotiables and
someone like Brian who was a strong and persuasive champion of design from
the outset.

From Accenture's point of view, it was also important that they protect and
also enable the design-led culture they were bringing to Accenture. Being
able to tap into the design capabilities offered by Fjord has changed what
Accenture can do for their clients. The most immediate benefit can be seen
in how they are helping their clients innovate, be more human-centred and
manage change better. However, the longer term and more significant benefit
to Accenture is using design to help it become a more human company.
Accenture has always been very client-focused, however their business and
technical functions have always been the dominant area, driving their culture,
while design remained in the background. By bringing in design to lead the
drive to humanise the organisation, they are tapping into design's role as the
Humaniser.

Why is this important to Accenture? First, they know that the millennial
generation is quite different from the generations before in terms of their
expectations about their work environment and how much they work. They
also place much more value on having a good work life balance. While
Accenture has a reputation of being able to implement large business
transformation programs and technological systems efficiently and
professionally, they realised that the ability to continually innovate requires

a change in culture to enable a more creative culture to flourish inside Accenture. In order to do this, they are enacting design in a *Humaniser* role to help employees focus on the human dimension for client work, but also as a *Cultural Catalyst* to build a creative culture within Accenture.

'We can no longer rely on an industrial model where you create an efficiently uniformed product pushed through controlled channels. Instead, we are working with clients to tackle very challenging problems. For that you need an organisation that's as flat as possible, far away from a hierarchical organisation. It should not obsess with structure and hierarchy, but instead focus on collaboration and communication. And without a positive and vibrant culture it's very difficult to have highly productive self-directing and efficient teams who drive at amazing outcomes for a number of different clients across industries. So I think that a culture of trust, transparency, openness and feedback is a must-have in order to be successful with the best businesses.'
Olof Schybergson, CEO, Fjord

Signs of culture change in Accenture

Fjord has been part of Accenture for the past three years and culturally, things are beginning to change in Accenture. Evidence of this change can be seen in a number of ways. At a more day-to-day level, the way Accenture management communicate with employees has changed significantly. Prior to Fjord joining, email communication was rather formal and often didn't give a sense that the employee was at the centre of the communication. It was about 'what's happening' rather than 'what's happening and what it means to you'. Communication has become much more visual, personal and directed.

'It's just a more exciting and personal way of communicating with everybody. We now use a lot more videos in our communication and they are visually more appealing. I give Fjord a lot of credit for that. Although they're not the ones designing the communications, I remember two years ago when we saw a Fjord presentation and everyone would be like wow that feels fun, cool and interesting. I want to communicate like that.'
Shelly Swanback, COO, Accenture Digital

At a more strategic level, Accenture has completely changed the way they manage performance internally. They have eliminated rankings and annual performance reviews for their 360,000 employees since autumn 2015 and instead focused on putting the person at the centre of the review. This shift made headlines when it was first introduced and has been widely reported in the business press. The new system is a much more fluid system designed so

What are the conditions for impact in Accenture & Fjord?

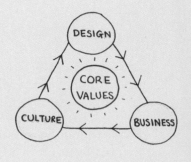

- Establishing cultural boundaries from the beginning.
- Respecting each other's cultures, expertise and capabilities.
- Protecting a design-led culture cultivated in Fjord whilst still aligning common values across both organisations.
- Close collaborations to ensure an osmosis effect between the two cultures.

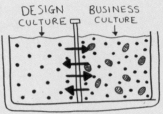

What have been the challenges so far?

- Bringing two very different cultures together.
- Establishing trust and cultural boundaries–what should be protected and what should evolve.

What type of change still needs to be achieved?

- Growing Fjord's role and influence in Accenture without sacrificing quality and identity.
- Fully leveraging design capabilities to support change in Accenture and with their clients.

that employees receive timely feedback from their managers on an ongoing basis following project completion. Fjord was instrumental in helping Accenture design the new process and tools to enable these conversations to happen and to focus on individual aspirations. This is another example where design is used in a *Humaniser* role. And while Accenture isn't the first large multinational corporation to move away from annual performance management (Microsoft and Deloitte did the same), Fjord has had a central role in influencing and leading some aspects of the work. This shift towards a more human organisation has also changed the type of talent Accenture attracts which is especially important considering that Accenture wants to inculcate a more creative way of working.

'It's less about the person's rating and instead more about the person's aspirations. Let's talk about the actions we need to take to help them grow, to help them be successful. It's just a completely different conversation and Fjord helped design the tools that we use and the way we're communicating about that.'
Shelly Swanback

Impact on Fjord

The changes taking place isn't aren't just happening at Accenture but are also having an impact on Fjord. On an individual level, the merger with Accenture has given Fjordians new opportunities: from a growth perspective there are new challenges to address, and from a career perspective there are a broader set of skills and roles to grow within Fjord but also potentially in Accenture. From a business perspective, Fjord has grown to four times the size they were prior to the acquisition, expanding their offices globally to 22 locations and operating in new markets such as in the Far East and Latin America. They are also able to leverage their sector expertise within Accenture and the network's access to global clients. Partnering with Accenture has also meant that a lot more projects come to fruition due to the vast resources and capabilities Accenture can offer in order to support clients during the implementation stage.

It's also important to note that the change roles have also made an impact at Fjord. While one of the early struggles has to been to maintain and protect their culture, the merger has also had a positive affect on Fjord. Design's role as a *Community Builder* has helped them build new communities and synergy with their Accenture colleagues. The hands-on approach to projects and its immediate impact on clients has helped win over colleagues' initial uncertainties about the design approach. Seeing the impact that design can have at a strategic scale has helped Fjord focus on the broader picture and not just on the designed outcome. Fjord realised how important it is to understand the business and what their goals are. This extends to how they now prepare for meetings. Adopting a common practice from their Accenture colleagues, they now allow more time to plan and understand who's in the room, what perspectives they have, and how to deal with them. Accenture has had a lot of experience doing this through their relationship and account building activities.

Difference between small 'd' and big 'D'

Bring part of Accenture has enabled Fjord to expand their practices and use of design beyond product and service innovation. When they were an independent design firm, they tended to attract more specific briefs where companies were looking for solutions, a design challenge or design opportunity–what you might call the small 'd' of designing. These kinds of projects were often very focused on the design component and involved a client who was very knowledgeable about design and acted as an internal champion for design. However their current work is often about the broader engagement where it's focused on a transformative change for an organisation–which could be called the big 'D' of designing. And while design is not the only aspect in this activity, it's role as *Cultural Catalyst* is crucial.

'One key learning for me has been the need for broader collaboration and embracing different skills sets and points of view. To truly put design at the heart of society, put design at the heart of the client's business, you do need to think of design as an amazing catalyst and a central driving force for change. You also need to push way beyond the traditional borders of design to be open to collaborate with people from many different perspectives, be it technical, be it change management professionals, industry experts, logistics, supply chain experts and so on to truly bring about the remarkable change that design can catalyse and often lead.'
Olof Schybergson

This type of transformational change also often does not occur without some initial difficult conversations with clients. Design is used in a *Friendly Challenger* role by the Accenture project teams to ask uncomfortable questions of their clients and to question basic assumptions that often go unchallenged. And while these types of activities make everyone uncomfortable, Accenture and eventually their clients see value in this process in order to help them identify new opportunities as well as improve on current practices. Design is used as a catalyst to help organisations change through its design methodologies and tools. It helps organisations understand what it is within their own structure that's stopping them from being able to respond to changes, how to unblock it and how to ignite these catalysts within their own organisations.

Another key learning that Olof has observed in terms of shifting their design practices from the small 'd' to the big 'D' is that, in order to have meaningful long-term sustainable change, it's important that the design vision is aligned to business goals and outcomes. This might mean improved employee engagement (as is the case for Accenture), helping an organisation better connect with a future customer or driving new sources of digital revenue. Ensuring that they work towards those outcomes is really important in order to make an impact through design.

Putting design at the heart of everything

The last three years have been an eye opener for both Fjord and Accenture. For Fjord, being part of Accenture has enabled them to bring their design capability to an even broader range of clients, sectors and countries than ever before. They joined at a time when the world has become much more open to design being used as a change agent and to putting design at the heart of organisations. For Accenture, having access to Fjord's design capabilities has not only changed the way they work with clients, Fjord's design-led culture is also helping them transform from a business and technological focused organisation to one that is more human, open and collaborative.

ZOOM Education for Life: Catalysing a shift in the Brazilian education sector

Delivering innovative learning solutions to a Brazilian market

Since 1996, ZOOM education for life based in Sao Paulo, Brazil has developed innovative learning solutions that range from product sets, educational booklets, tablets and teaching support materials and services. ZOOM has partnered with LEGO® Education since 1998 and has been a LEGO® partner in Brazil, serving more than 2 million children and young people with an active base of 150,000 students every year. Around 8,000 schools have used ZOOM's products, with an active base of about 450 schools. They also have 14 franchisees across Brazil.

ZOOM has been focusing on the private sector since 2015 due to increasing demand from the growing middle classes. ZOOM has several educational programs that are curricular and extracurricular for a broad target audience (from as young as 3 years old to adults) and are specifically targeting 21st-century skills through performance in the STEAM (Science, Technology, Engineering, Arts and Maths) subjects.

ZOOM has always strived to innovate in the education sector and their changing business model illustrates this. In 2003, they launched their ZOOM program and opened up their ZOOM education methodology for life to small schools as well as large educational networks. This provides educational establishments (especially private schools) a market differentiation for their students without requiring a huge amount of investment.

Throughout its history, ZOOM has always tried to exploit different market niches in order to find new competitive advantages and ensure

Who we spoke to
Erick Augusto Moutinho, Product and Innovation Director
Victor Barros, President

Why change?

A design driven-innovation process was brought to ZOOM because they wanted to revolutionise their product development process to help them exploit the challenges and opportunities in the Brazilian education sector, which is experiencing increased investment. They want to extend learning beyond the classroom, allowing students to take more ownership of their learning and challenging the fundamental relationship between student and teacher. The innovation process has led ZOOM to develop a complete learning ecosystem with a well-defined journey through a digital platform. Shifting focus from products and services to an ecosystem-centred business model has put the customer/user at the centre of their solution and given them a strong foundation for how and why they innovate.

Design roles that enabled change in ZOOM

Types of changes achieved through design

Since 2014

Changing products & services

Changing organisation

What has a design-driven approach brought to ZOOM?

- Products that meet the markets demands.
- Increased profits.
- Helped them anticipate challenges in the market.

the profitability of the company by developing their own products, services and educational materials. Although ZOOM started life as a distributor of education products, their range of solutions now includes support in the form of a package and contract format, similar to a telecommunications company. They have moved from being a distributor of products to a company that offers a full end-to-end service for the educational sector.

This move to a digital platform that underpins their products is evident in their four main business areas. ZOOM describes their solutions based on four pillars. Although ZOOM sells products from LEGO® Education, around 85 per cent of their revenue comes from their specially created didactic solutions and curriculum that has been designed to work with LEGO® Education sets. The second pillar includes the 1-3 year support programmes they offer to schools in the form of pedagogical and technological advice to schools. This support helps the schools make the most use of the products offered by ZOOM. Teaching material consists of educational booklets and services related to a set of products that enables putting into practice the proposed activities.

The final pillar is a digital platform where students can access and develop content, including learning through tablets in as well as out of the classroom. This last pillar brings a huge added value to all stakeholders and allows the development of a unique set of ZOOM´s solutions.

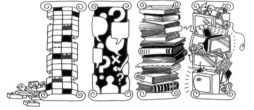

Education sector in Brazil and Latin America

To understand the transformational journey undertaken by ZOOM, it is important to first understand the Brazilian educational context.

Brazil has the ninth largest economy in the world based on nominal GDP. And yet its education standards still lag far behind. In 2000, only half of Brazilian children finished primary education according to the OECD sponsored Programme for International Student Assessment (PISA). Out of the 31 countries who took part in the assessment, Brazil came out last in all areas of assessment; reading, mathematical and scientific literacy. This was a dire assessment for a country with ambition to become a fully developed country.

Improving education in any county takes time. In Brazil, this challenge is further compounded by difficulties in addressing vested interests of administrators, overcoming a strong teachers' union and reforming an outdated education system. However, the shock results in the OECD assessment prompted the incumbent and subsequent governments to fully commit to improving the standard of education in Brazil. As a result, Brazil has steadily increased public spending on education from 10.5 per cent of total public expenditure in 2000 to 19 per cent in 2011. This is well above the OECD

Stages of transformation

1. The creation of the Innovation and Product Management team.
2. Developing ZOOM's innovation model.
3. Trialling the model through new products (ZOOM Technology Education, ZOOM Early Childhood Education, ZOOM Engineering.
4. Looking for partners to grow its eco-systems to deliver educational material through a digital platform.

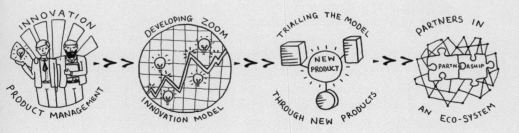

What can we learn from ZOOM's story?

A well-developed innovation process can bring immediate value to an organisation.

Change happens quicker in a smaller organisation.

average of 13 per cent and is the fourth highest among all OECD and partner countries with available data. In the last 2012 PISA assessment, Brazil was ranked 58 out of 65 countries. And while significant improvements have yet to be achieved, increased investments and efforts have significantly accelerated this progress.

In 2014, the Brazilian government signed the ten-year National Educational Plan (NEP) and one of the key goals of the plan is to invest ten per cent of the country's GDP in education by 2024. In addition to the current 6.4 per cent GDP invested in education, a law passed in 2013 earmarked 75 per cent of petroleum royalties and 50 per cent of all sub-salt layer oil royalties for education. One of the key differences between the NEP and the previous education plan is that it is a constitutional decree, signed into law. This means the planned investment aimed at ensuring the free or subsidised vocational and higher education will not be jeopardised by a change of government. The NEP also establishes the expansion of programs such as FIES and FIES Technician (Student Financing Fund–Higher Education and Technical) and PROUNI (University for All Program), as well as expansion of enrolment in professional education with an increased offering of courses in private institutions. These significant investments have opened up a lot more possibilities for ZOOM since they are already well established within the education sector.

Developing an innovation process

Although ZOOM has an established and profitable business, they are acutely aware that in order to really capitalise on the increased investment in education and to meet these current challenges, they needed to significantly improve their product development process. They have always relied on insider knowledge from educational professionals to drive their product and service development. While this approach worked to a certain extent, ZOOM found that this method tended to converge on ideas with a predefined solution rather than offering any new insights and opportunities in the learning environment.

ZOOM decided to introduce a design-led innovation process to help them transform how they work and think in a structured but differentiated way. They wanted to bring the customer/user to the centre of the solution. They were also keen to expand learning beyond the classroom. They wanted to involve all stakeholders in the education process, for example the educators, schools and students themselves. They wanted to move towards offering a complete educational ecosystem with a well-defined learning journey. One of the key changes to their business model was to develop and make central a digital platform for their products. To do this, they had to consider the interactions between the offline and online world.

The process of developing an innovation model and using that platform to drive change inside ZOOM started in 2014. Erick Augusto Moutinho was invited to join ZOOM to lead their innovation and product development team set up as part of their five-year plan. Victor Barros (ZOOM's president) was instrumental in getting Erick involved. Having come from a tech start-up culture, Victor wanted to build a new way of thinking and innovating in ZOOM.

The innovation and product management area is responsible for mapping trends, identifying new market opportunities, developing new products using an innovation methodology and managing the products already launched. They now consist of 20 people divided into innovation, product design and digital platform teams. The digital platform team was started two years ago as ZOOM started to expand their business focus and now has teams dedicated to UX and UI, development and quality assurance.

When Erick joined, ZOOM was just starting a strategic project called ZOOM Technology Education (ZET). The ZET project marked the start of a new design process for ZOOM. Erick and his team experimented with a number of innovation models and approaches to see what worked and what did not work. At that point ZOOM was shifting its business model from being a product distributor to a company that publishes new materials as well as being a digital service provider. They were not only developing physical and digital educational materials, they also acted as a technology partner to schools.

Erick and his team started developing ZOOM's innovation model by studying 16 innovation models from around the world and mapping them to their education market. They created a partial model and tested it with staff. They wanted to find out how the model was received in ZOOM and whether it would be suited to ZOOM's culture. They tried to understand how people and departments regarded this model and whether it was fulfilling its organisational needs. They continually refine their model into what they now call ZOOM's 'Innovation Funnel'.

A framework for innovation

There are four stages to ZOOM's innovation funnel. And while it does not differ significantly from existing innovation models, the focus on starting with challenges to identify opportunities and co-creating with users is a significant departure for ZOOM's culture.

The first stage–'Inspiration' is where new opportunities in the form of 'challenges' are explored through research and interviews with potential users of the solution. After defining the challenge, the product and innovation team start to immerse themselves in the opportunity space through the usual

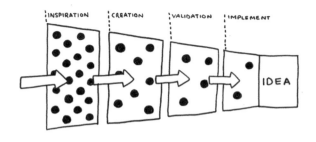

design research tools like shadowing, in-depth interviews, CDA matrix (certainties, doubts and assumptions) and cultural probes. This process enables them to empathise with their users and to immerse themselves in the challenge in order to understand all aspects in detail. As the end of this first phase, they turn the co-creation process into insights and opportunities. This first stage is really important for ZOOM, not only does it mark a significant change in approach for ZOOM, it's a departure culturally from Brazilian businesses.

Starting a project without a clear answer is seen to be a risky approach by Brazilian standards. Instead, they want to see returns on investment quickly and focusing on clear and existing problems seems the best way to achieve this. So it was actually quite a gamble for ZOOM to sanction a process that was untested. However, adopting this approach has certainly yielded results and importantly has given them an edge over their competitors.

Once a challenge has been identified and opportunities mapped to it, the next 'Creation' stage starts the process of co-creation of solutions with users. The aim of this process is to generate a diverse number of ideas and so the broader the range of user profiles, the better. This means talking to educators, schools, students, parents and guardians. At the end of this phase Erick and his team will have identified a number of solutions that will be prototyped and tested.

'Having ideas is only the start. The key challenge we have is to convince people inside the organisation that the value of the co-creation stage is in the unexpected ideas that emerge. Yes we have two ideas that we knew from the start, but we also ended up with eight new ideas.'
Erick Augusto Moutinho, Product and Innovation Director

At the 'Validation' stage the co-created solutions developed in the previous stage are taken through their paces. It's important at this point to have a working prototype for users to experience all facets of the solution and give their feedback freely. The interaction is continuous and uninterrupted at this stage until the market, customers and users have validated the ideas. At this stage, the innovation team starts to develop relevant business models. The final 'implementation' stage starts looking at the various aspects of pricing, marketing, contracts, supply chain and other areas in order plan a route to market.

What are the conditions for impact in ZOOM?

- Clear evidence of what design brings to the business—in this case the higher sales figures for the products developed using this new innovation process.
- Developing an innovation process suited to the organisation and the sector.

What have been the challenges so far?

- Overcoming professional snobbery in the education sector.
- Convincing staff internally about the benefit of co-creation and asking the user.

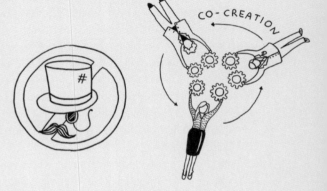

What type of change still needs to be achieved?

- Growing the innovation and product team

Learning to innovate

It has been a learning experience for Erick and his team to develop a model that suited ZOOM's culture and business model. As ZOOM offers both physical and online products and works as a publisher as well as a digital service provider, they needed to find a design process suited to how they work and what they offer.

The current iteration of the model has been developed through three new products, ZET (ZOOM Technology Education), ZEI (ZOOM Early Childhood Education) and ZEP (ZOOM Engineering). These three projects also represented how understanding of design has developed in ZOOM. There has been significant learning and adaptation in each of the projects for both the product managers and the organisation in order to understand the process of innovation and new product development. They initially started with a more flexible model and have now refined it to a more structured and streamlined process that is more suited to how they work. In general a new product takes between three to six months with three people dedicated to its development from start to finish.

The product development for ZET had already started before the innovation model was fully developed. So Erick and his team had to 'catch-up' with its development process and this meant that they were short of time to fully plan out the process. Although the product was a success when it was launched, the team felt rushed and they learnt that they needed to balance the planning of the project with its execution. They also realised that they needed to spend more time understanding the users and gaining more in-depth insights. Using this learning, their second product aimed at early childhood development was given more time for development and user research. And while they invested more time developing a better understanding of their users, they wanted to go a step further and use these insights to challenge pre-existing notions about education.

'In the first product we were fairly familiar with the market and were able to create a product that met its needs without needing to spend too much time in the development process. However with the third product, even though it was a new area for us, we were confident we were launching a product that had at least a ninety per cent chance of succeeding based

on the process we went through and the research, experimentation and feedback we received from the users and market.'
Victor Barros, President

In developing their next product, called ZEP, they started by understanding challenges and problems faced by users in order to establish a design challenge. They did not just make assumptions, they found out if they were correct or not. They devoted more time to the inspiration and the creation stages, which gave them more confidence going into the validation and implementation stages. Since the most time resource intensive phase is the implementation stage, getting the first three phases right significantly reduces cost and time.

Cultural Catalyst for ZOOM and the educational sector
'The culture changes in the most important point for us because in education we have a huge challenge to help our team think differently from the way that they are used to thinking.'
Victor Barros

There is a prevailing air of professional snobbery amongst the educational establishment and professionals, especially in Brazil. There is an assumption that educators know what is best for students and as a result, education and learning tends to be a one-way street. The simple act of focusing on users and understanding their needs has challenged the notion of the expert in Brazilian education. For ZOOM this was the biggest challenge to overcome.

ZOOM has an educational department made up of professional educators who are responsible for the teaching methodology, teaching materials and also for providing educational services to ZOOM's customers. In a culture where teachers are expected to be an authority on educational matters, it has been challenging for ZOOM's team to adopt a more user-focused approach and work collaboratively with students, parents and other educators. However by stimulating the internal culture to change through a clear focus on peoples' needs and deep empathy as a means of approaching sensitive, cultural challenges the product development team was able overcome resistance to change. Not only was the education team able to relate user insights to the ideas that were generated, it gave them many more new ideas that were not obvious to them at that time. It also helped that the first product created using this new design-led innovation focus was a commercial success and this further cemented the approach for ZOOM.

It was also a challenge introducing unfamiliar design methods (like shadowing and observations) to staff, but once they realised how it could help them gain new insights and offer more directive product development, they were quick to adopt it. They recognised that design is not just a final add-on at the end of a product development stage, it is a mindset that helps them empathise with both internal and external users and has steered their focus from a pre-defined solution to uncovering challenges that need to be solved. This change in practice has not only affected internal staff but also influenced the practices and attitudes of the educational partners and schools that work with ZOOM.

Offering a framework to deliver success

At the leadership level, the explicit design-led innovation process has helped the executive team overcome the fear of the unknown and uncertainty by providing an explicit and clear process as a reference point. As a result, the executive team is much happier to deal with uncertainties at the beginning of projects and have bought into the idea that the more time spent at the front end of the innovation process, the less time and resources are needed to develop a successful product. Although vastly simplified, the innovation model provides a clear roadmap from start to finish, providing a psychological safety net for people in the organisation not accustomed to exploratory and divergent thinking and acting.

In all the cases we have featured in the book, confidence in following a design process is only cemented by tangible commercial success. In the case of ZOOM, the immediate success of the three new products helped the innovation team prove their worth. The executive team's confidence in the process has provided the innovation team with added autonomy in relation to decision making during the course of projects. The team is now trusted to deliver success.

Linked to this success, the attitude in ZOOM towards innovation has changed. When the innovation team was set up, they were initially only involved in analysing and bringing in new ways of developing products. They were mostly involved at the front end of product development. Additionally, the product development process was also shared between three to five departments. However this has now changed and all product function is centralised in the innovation and product management department. As a result Erick is now responsible for the end-to-end product development in ZOOM.

Another visible sign that the culture has changed in ZOOM is the development of the ZOOM's prototype lab. Erick is setting up a prototype lab in order to test how their solutions would work across different devices. One of the key challenges in Brazilian schools is Internet connectivity and to help resolve this ZOOM has started providing schools with Wi-Fi solutions so they are able to use ZOOM's digital platform. The lab will also be used as a user research lab, enabling ZOOM to test ideas with users, offering a different user environment outside of schools.

Challenges ahead

Despite the significant change brought to ZOOM, the Innovation and Product Management team driving the innovation process is actually quite small. Erick oversees the innovation process and is aided by only one other staff member. Each new product is assigned a product manager and usually only 4-5 people are involved in the development process directly. In order to solidify the change in culture and to continually support this way of working, it is important that the capacity of the innovation team grows.

Although the education market in Brazil is set to grow with additional investment as promised through the NEP plan, the Brazilian economy has also stagnated in recent years and ZOOM is being mindful that they need to keep innovating to survive this current downturn.

'We need to think more lean; we need to balance between the short and the long term. We have to invest and ensure we maintain a culture of innovative thinking in the company to survive through this period.'
Erick Augusto Moutinho

In order to stay ahead of the market, ZOOM will have to be continually creative in its product development strategy. Erick refers to the s-curve of a technology adoption life cycle. The starting point of the S (if viewed facing down) illustrates the stage when the product is in development while the top of the S illustrates the product at its peak adoption. The end point of the S illustrates the end of the product cycle. ZOOM has products at different stages of maturity in the market. Some products will be at the start of their development cycle, some in the middle while some might have already reached a saturation point and be at the end of their product cycle. Being innovative does not necessarily mean constantly coming up with new products but to also looking at ways to create add-ons to existing products that will enhance and prolong the product's lifespan. ZOOM is also looking to leverage existing solutions like the Google Cardboard which turns any smart phone into a VR headset by developing products that can be used in conjunction with it.

ZOOM's business model has changed significantly from where they started 20 years ago. As part of their expansion into being a digital provider, they are positioning themselves more as a hub that brings people (teachers, students, parents) together. Like many of the other organisations we have featured in the book, they are looking at building partnerships with other organisations in order to build an eco-system of products and services that offers a comprehensive experience for their users. They are really attempting to turn education in Brazil on its head and change the way learning is experienced using design as its key catalyst.

Deloitte Australia: A leading innovator in the professional business service sector

Deloitte Australia and Design

The story of how design is becoming a key capability in Deloitte is richly demonstrated in Deloitte Australia. A number of factors led the executive team at Deloitte Australia to rethink its position and strategy in what was considered a very saturated market in the early 2000s.

In a sense, Deloitte Australia was an early adopter of design and pre-empted a trend in global accounting firms over the last few years to expand their consultancy activities by offering design services in an effort to offer a full range of services to clients seeking complete digital solutions. This has often been done through acquisition of existing design and digital agencies, for example Ernst & Young acquired Seren (a digital design agency) in 2015, KMPG acquired Cynergy Systems in 2015, PWC bolstered their digital design capability through the acquisition of Ant's Eye View, Intunity and Optimal Experience over the last few years.

Although efforts by Deloitte Australia to use design strategically started as early as 2003, it only really gained momentum after 2008. With a new CEO, Giam Swiegers, in place in 2003, Deloitte Australia realised that the only way to create new and sustainable competitive advantages was to innovate in a quick and agile way. Deloitte's leadership wanted to leverage the firm's traditions of solution-driven expertise with a willingness to be open, creative

Who we spoke to
Shane Currey, Partner, Design For Business and Narrative Strategy
Leon Doyle, Partner & Head of Experience Design at Deloitte Digital
Frank Farrall, Lead Partner, Deloitte Digital, Asia-Pacific
Jo Rhoden, Partner, Design Lead, Strategic Capability and A&A

Why change?

Deloitte Australia decided that the only way to create a new and sustainable competitive advantage in the professional services sector was to adopt a more humanistic and insight-led approach. It wanted to leverage the firm's tradition of solution-driven expertise with a willingness to be open, creative and exploratory. This enables the people within the business to work more collaboratively with clients, identify opportunities earlier and deliver targeted solutions that respond to clients' needs rather than assuming a pre-defined solution.

Design roles that enabled change in Deloitte Australia

Types of changes achieved through design

Since 2008

Changing products & services

Changing organisation

What has a design-driven approach brought to Deloitte Australia?

- Differentiation with clients and proof that it works.
- A more human-centric approach.
- Enabled contemporary workplace practices that allow staff to work collaboratively with clients to identify problems and work towards finding a solution.
- Created different and meaningful outputs and outcomes.

and exploratory. One of the ways they thought they could do this was to buy-in the expertise. The acquisition of the Eclipse Group is a good example of this.

The Eclipse Group was acquired by Deloitte Australia in 2003. Eclipse was one of the few sizeable web development companies that survived the dot com crash. In the early days, Eclipse was run independently from Deloitte Australia and, if it were not for the foresight and leadership of certain key people at Deloitte Australia, design would have remained on the sidelines. The change started in 2008 when Deloitte Australia embarked on the journey to transform itself from a traditional professional services firm to a bold innovator in the sector. Part of that journey involved the key decision to integrate Eclipse into the consulting practice within Deloitte Australia. It was rebranded as Deloitte Online. Fairly soon after this, another entity, Deloitte Digital, was set up to deliver professional services online. At that time, a majority of the professional services firms had not embraced digital and were still delivering services face to face. Deloitte Digital's aim was to pioneer the delivery of professional services online and create a range of new online services covering Education, Innovation, Compliance, Surveys and Benchmarking. In 2012, Deloitte Digital went global by establishing nine digital studio offices in the UK, US and Australia. It was also the right time to merge Deloitte Online and Deloitte Digital, since the teams had always worked closely together and shared not only the same space but also a design-centric culture.

Deloitte Digital is now the largest design group in Deloitte Australia. Frank Farrall is the lead partner for Deloitte Digital Asia-Pacific and Leon Doyle is the Head of the User Experience team, which itself consists of the largest concentration of designers (over 200) in Deloitte. They are a full service digital agency and provide strategy services, design as well as large digital transformation programmes for clients.

'We are arguably the largest design practice in the country and almost certainly the most impactful in the country.'
Leon Doyle, Partner & Head of Experience Design at Deloitte Digital

Although their range of activities is still primarily focused on digital design, they are increasingly acquiring a range of designers and design agencies with specialist skills in service, strategy, motion, interiors and architecture, with a recent acquisition of brand and spatial agency, MashUp and digital storytelling consultancy, The Explainers. They are moving into the intersection between digital and physical design and exploring how that changes the narrative and the conversation in different spaces such as in retail and banking.

Stages of transformation

1. Buying into design capabilities and offering them as a service externally through the Eclipse Group in 2003. Focused primarily on web design.

2. Awareness of design thinking and buy-in from executive team starting from 2008. Bringing the Eclipse Group into Deloitte Australia.

3. Hiring Maureen Thurston as Principal of Design Leverage (an explicit Design Leader role). She led the 'Different by Design' programme aimed at leveraging the power of design in business.

4. Setting up the Design for Business team (originally the Design, Visualisation and Storytelling team) in 2014 to support consulting teams with external clients. At the same time in 2014 – developing the Deloitte Business Designer program, building on Maureen Thurston's work piloted within the Assurance & Advisory practice and then taken more broadly across the firm.

5. Rationalisation of the design functions across Deloitte Australia and developing a strategic plan for the use and role of design in Deloitte.

What can we learn from Deloitte Australia's story?

Design functions can emerge in an organisation through a number of routes and engage with the different aspects of the organisation in different ways.

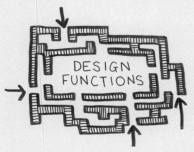

Design functions in organisations are complex and tend to grow organically. However for it to have sustained impact in the organisation, design needs to be consolidated formally and become a core part of the corporate strategy.

We need inspiring leaders in the position of CEO as well as at the team leader level to maintain momentum and enact positive changes.

Serendipitous encounters

We know that using design to transform an organisation does not just happen overnight but requires a concerted effort as shown by our examples in this book. For Deloitte Australia, several things occurred. The then-CEO Giam Swiegers met leading design thinker and author Roger Martin in Sydney in 2010. A few weeks later chief strategy officer (CSO), Gerhard Vorster, bumped into Roberto Verganti, another pioneer in design thinking, in London. The Sydney managing partner John Meacock, and current CSO, read an article on design by Sara Beckman and Michael Barry, professors from Berkeley University and Stanford University respectively. Not long after, Verganti was invited to run a two-day design workshop focused on Deloitte's new Risk Services strategy with Vorster, Swiegers and other key members of the executive team. These serendipitous encounters became the seeds that eventually led Deloitte Australia to make a commitment in 2011 to use design thinking to change its own approach and operations in order to redefine the way professional services are delivered.

This commitment was followed through in a number of ways. In 2012, the entire executive team attended a course at Stanford's d.school. They were introduced to design thinking, tools and methods, and importantly challenged to think like a designer. These sessions, run by Beckman and Barry, were delivered to hundreds of Deloitte's leaders through a design thinking program. The establishment of design internally in Deloitte Australia can also be attributed to the efforts of Maureen Thurston. Maureen, an industrial designer, educator, consultant and entrepreneur was hired in 2012 as the firm's Principal of Design Leverage to lead a unique initiative that deployed design as a catalyst for change; bridging the gap between creative capability, business strategy and top-line growth. Maureen created the 'Different by Design' programme that has led to Deloitte undertaking a number of internal projects and initiatives using design, and she created differentiated client experiences supported by design thinking.

Deloitte Business Designer programme

A number of key things emerged from this concerted effort to embed design in Deloitte Australia. Building on Maureen's work, the Deloitte Business Designer programme was created in 2014 and piloted within the Assurance & Advisory (A&A) practice. Jo Rhoden at that time was the programme director for Audit Differentiation and started working with Maureen from 2012. During this time, she quickly understood the power of design in re-shaping how professional services were experienced. Design was the 'how' that enabled other differentiating technologies such as data and digital to become meaningful and actionable for clients and teams. This was the catalyst for creating the Deloitte Business Designer programme, which was developed and tested within A&A before being offered more broadly across the heritage businesses of the firm as part of Jo's current role: leading design within the Strategic Capabilities team.

In order to embed design within the Strategic Capabilities team, Jo runs a three-tier programme. Tier One builds from Maureen's work and is to help build understanding and advocacy; to help people understand what design is and how it can be used in the context of their work. Tier Two is project focused, using specific team-based projects as an opportunity to learn and apply design. Finally, Tier Three is focused at the transformational level of empowering practitioners, from any of the heritage services across the business, with deep design skills so that they can apply design to their work and, importantly, lead their teams in learning and realising the value of a design-led approach. This is the Business Designer programme. It's the most involved and hands-on of the three tiers and it involves 3 weeks of immersion, learning and applying design, weaving it together with data and digital as well as applying it to specific, pre-agreed client engagements. This is followed by a 6-month consolidation period back in the business, working with engagement teams before coming back for a final 2 weeks of learning and refining skills and approaches. It's intensive, practical and supportive and its aim is to build confidence in those going through the program so that they can take their team(s) on the journey to doing their work differently by design.

Design for Business

At around the same time when the Business Designer programme was initiated, Deloitte established a small and agile team that was originally called Design, Visualisation and Storytelling. The team assists Deloitte Consulting to solve their client's problems using design, visualisation and storytelling approaches. The Design, Visualisation and Storytelling team originally described their work in three areas[1]. Firstly, design thinking is used to facilitate design-led problem solving guided by the Deloitte Design Process. Visualisation is the second practice used to help clients communicate, collaborate and engage the senses. And finally, storytelling is used to create and tell stories in order to bring meaning to the work as well as activate people's imagination.

The DVS team changed their name to Design for Business (DFB) in early 2016 to more closely reflect their current work and future aspirations. Although their core practices have not changed significantly, the focus has been realigned to encompass design-led innovation, design capability and narrative strategy. When we interviewed Shane Currey, who was instrumental in the creation of the original team and is still currently leading it, he felt the name change was important. The 'Design for Business' name moves the focus away from specific design skills to a name that is more inclusive and representative of what they currently do. The team was created to allow consultants to have access to designers on their projects. The new name is also more representative since it's about using the design capability of visualisation and storytelling to help the business innovate. The current team is now 12 strong and its aim has always been to offer consultants access to design capabilities, which they can leverage for their projects.

Cultural catalyst from within

Often one of the key ways design is used to change an organisation is to act as a catalysing force. In the case of Deloitte Australia there have been different ways in which design has been used to bring about changes in practices. The Design for Business team collaborates with different teams in the consulting business, with a focus on key strategic clients. This operating model has two advantages. Firstly, it allows design and designers to add value across a wide range of projects. Secondly, it means that many more consultants are exposed to design and this has resulted in the embedding of design in the business. As a result, design is becoming a key component of Consulting projects.

In a similar way, but focused on a different area, Jo Rhoden pioneered the approach of developing design capability in addition to deep accounting and auditing skills within the Assurance and Advisory (A&A) practice. It's about turning the focus from process to people. This can be challenging at certain organisational layers where the work tends to be very process and compliance based. Compliance is a critical component of the work done within highly regulated industries such as audit and so the challenge is how to ensure quality and compliance is exceptional whilst still enabling the human-centred element. One approach Jo has been taking is to help analysts open up their curiosity and pursue good questions, rather than simply seeking responses. The ability to ask and follow good questions (using design in a *Friendly Challenger* role) is an incredible 'cross-over' skill which underpins not just the mindset of professional scepticism (a key attribute of a good auditor) but is also a core skill of the design approach and helps clients and advisors to understand their problems better, in order to find improved outcomes.

One of the challenges of embedding design approaches within heritage services is to open up people's minds to the possibilities it can bring to their own day-to-day work. This is best done by showing people and teams how to apply it in practice to audit, or advisory services or whatever the service offering is. Design needs to be interpreted and translated into the language and application of people at the front line of these service offerings. As part of the programme that Jo has developed, she not only provides design training and skills

What are the conditions for impact in Deloitte Australia?

- Support from executive team (buy-in and make part of the corporate strategy).
- Having the right individuals to lead the design functions.
- Working with early adopters who are happy to try a different way of working.
- Proving design is profitable to the business.
- Creating a critical mass of designers that is more visible and has presence in order to start influencing culture.

- Offering the right support and environment to keep designers interested and motivated.

What have been the challenges so far?

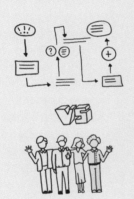

- Moving from a culture of finding answers to questions to one that starts with asking the right question.
- Moving from a process-led culture to one that combines the process and human-led approach.
- Moving from a siloed way of working to working across teams.

What type of change still needs to be achieved?

- To introduce and use design consistently across the different areas of Deloitte Australia.
- To bring together and rationalise the different design functions and activities across the organisation.
- To cohere design knowledge currently residing in different teams in order to share good practices.
- To embed design as 'the way we do things around here' within heritage businesses.

to complement existing professional skills, but also helps individuals develop the confidence to challenge the current systems and ways of doing things to ultimately bring deeper value to their teams and clients.

Brokering new ways of working

Another change in practice that has created an observable change within the consulting practice is the movement away from a siloed way of working. This philosophy is very much reflected in the way the Design for Business team was set up. The consulting business has different function areas such as human capital, technology, customer and digital and strategy operations. DFB was set up to bridge all functions and is sponsored by the national executive. It sits outside of any of the main competencies and is not 'owned' by any area. It means anyone is free to approach and consult with them. This also removes the tension between the different teams since there is always an underlying pressure to bring in a projected amount of revenue for each team. Revenue generated through the work of the DFB team goes directly back to the team. By sitting outside and offering their services as a neutral partner, the DFB team is going against all the normal conventions of how consulting teams have worked in the past. It was a strategic move that is aimed at activating design across the business. The role of design in this example is not only acting as a *Cultural Catalyst*, but also as a *Power Broker*, being able to leverage its 'neutral' stance and its customer focus to enable the consulting teams to be a more fluid and flexible in terms of how they work and use internal resources. The inherent human-centredness of design helped neutralise power dynamics at play enabling a more cooperative rather than a competitive culture.

Working in a designerly way

Another observable change in the practice, especially within the consulting business, is the increasing visibility of visualisation of ideas and concepts in project spaces. There are more prototyping activities as well as visual reminders of customers' needs and motivations, which are positive signs of a human-centred approach to their work. Additionally, the culture is becoming much more activity-based and many of the walls have been converted into whiteboard spaces and soft furnishings have replaced traditional task chairs that were more formal and fixed. Shane Currey talks about this shift as representative of new contemporary workplace practices. The new generation of professionals do not sit at fixed desks anymore but instead move around, work in teams and are more fluid in terms of how they see their work. It is moving towards a more organic workplace as well as having a more human-centred focus in their work.

Conditions that made change by design happen
Support from the Executive Team

Using design to transform an organisation takes time, effort and a strategic plan. Although awareness of design and its eventual sponsorship happened through a series of serendipitous events, a clear strategic plan and support from the leadership team has been key to the success of using design in Deloitte Australia. Support from the leadership team has been by far the most important prerequisite to have in place before an organisation embarks on a journey of transformation using design. Not only is it important for the leadership team to 'get' design, they have to ensure that they provide the resources, environment and structure to enable teams like the Design for Business and Strategic Capability to flourish.

Being selective

At a team level, Shane has been very targeted in terms of whom he wanted to work with from the start. Having worked in Deloitte Australia for over nine years before setting up the DFB team, he had already built up a network of colleagues that he knew would be open to trying a different approach. This strategy enabled him and his team to build up a body of work that had demonstrable impact.

'Having someone in the business that has the networks, credentials and the respect to enable you to go and say to people, "try this" is really important in building the credibility of the team.'
Shane Currey, Partner, Design For Business and Narrative Strategy

Currently, at this more established point, the DFB team applies a prioritisation method not simply for strategic reasons but also to help filter through an increasingly high number of requests from teams. They prioritise projects they feel design could have the most impact on. Examples of projects that they have worked on included a project with Australia's leading bank, helping them redesign the mechanism in which people were able to donate in response to a natural disaster. Another example of a challenging project involved working with the human capital consultants and their client (a public sector organisation) to manage a change in leadership style and to help remove some undesirable, culturally-ingrained behaviours.

Similarly with the team of business designers now operating within the A&A practice, a framework is applied to ensure that this team of highly in demand individuals work on the business's most strategic priorities through the four lenses of: new pursuits, retention, internal initiatives and capability development.

Easy to understand process and adaptable tools

With the adoption of any new approach to an established practice, it's important that the process described is not only easy to understand, but also transparent and can quickly show results. One of the benefits of design is its collaborative and practical approach to problem solving that strikes the right balance between design, innovation, pragmatism and simplicity. The Design for Business team has developed a design process that is suited to the way Deloitte works. The team uses a whole suite of design tools, but like other teams we featured in our book, tools are only a means to an end and what is far more important is to support the teams to stay true to the process and mindset of design by focusing on user needs. The DFB team used design as a *Framework Maker* to give them and the teams they work with the confidence that they could achieve their goals and ambitions using an easy to understand process adapted to their specific needs.

'We have created a Deloitte design process to help us educate people about what design is in a short period of time but also to ensure we have consistency in our communications internally and externally. However in reality we would use different models and different frameworks from different places depending on the problems. We are much more flexible in terms of using different ways of looking and using design.'
Shane Currey

Building critical mass in design

In many examples in the book, there is usually a focal point where design functions and capability reside. Both Leon Doyle and Shane Currey agreed that it would have been difficult to start building a design-led culture without first achieving a critical mass of people who were already working in that way. While Deloitte Digital's team has predominantly come from Eclipse employees, who were mostly designers; Shane's DFB team had to be built from scratch. In this case, it was important for Shane to ensure he got the right design talent into his team by creating a compelling career prospect in a sector that isn't naturally attractive to designers. Not only was it important to recruit the right type of designers, it was also important to have the resources to reach critical mass by building their own community of practice. Additionally, in Jo's work she has created design careers for technical professionals from heritage services where previously no such career route existed. These are entirely new roles that have not been imagined previously in the context of these services.

'When I tried to set up these design teams in other places across Australia they failed because one person doesn't make it a team. If you sit a single designer amongst consultants it's very hard for them to break through the status quo. When I proposed to set up the Design for Business team, I made sure we had critical mass immediately. I proposed at least five people for us to feel the momentum and make an impact that matters.'
Shane Currey

Building communities of practice

Not only is setting up the right team important, having a leader who is able to constantly energise and reinvigorate the team on an ongoing basis is equally essential. In that sense there isn't a huge difference between being a design or a non-design team leader. However the key difference is that a design team leader has to be sensitive to the specific needs and support required by creative people. It's not as simple as using the usual off-the-shelf training like to 'how to conduct a meeting effectively', but rather to source additional material that would be inspirational or insight-making for the team. For example, Shane brought in a storyteller that had worked on Hollywood scripts to explain to the team how to build a compelling story.

One of the interesting insights we found in Deloitte is the way they explicitly bring together the analytical and rigorous side of the business with the creative and insight-led approach. They have been very deliberate in meshing the two by 'twinning' leaders with different propensities to help build relations with people in order to find synergies and areas for design to work in.

'I, myself am twinned with three other partners in the firm who are much more left-brained than I am and it's about twinning, finding and building relations with people who are open to exploring a different way of working. I'm loving twinning with those guys, as I learn a lot more structure and rigour from them.'
Leon Doyle

Redesigning design at Deloitte

The story of design in Deloitte Australia illustrates how long it has taken for design to gain a foothold in a large organisation. Since 2008, Deloitte Australia has embarked on a transformational journey to become the leading innovator in the professional service sector. The way design has been brought in and expanded in Deloitte through Deloitte Digital, Design for Business and the Strategic Capability team has been organic up to this point. There are now over 300 designers and design thinkers in Deloitte Australia across the various teams. Although these different teams have various aims, focus

and expertise, their activities often overlap and they can be seen as peers who view themselves as agents of change by using design to improve organisations, whether it's for their clients or in their own organisation.

The different practices of these different teams are key to building design as an important function in Deloitte. Their next role is to create some structure to help the communities of practice rationalise their design functions and map out how design fits in the overall corporate strategy of Deloitte. The next step in the evolution of design in Deloitte is to bring these different communities of practice together with the aim to operationalise and create a formal structure to help the communities share best practices. It would mean bringing together existing approaches, methods and tools and to start building a design knowledge base.

To do this, Frank, Leon, Shane and Jo have been tasked by the executive team with designing the next iteration of design at Deloitte. They are part of a group called Design Stack and have been given the responsibility by their CEO to essentially redesign design and what it means for Deloitte. How will this be done practically? They have each been given an area of responsibility and focus. Frank is the group sponsor and he oversees the group's decisions. Jo is looking after the capability aspect within heritage services and building the right competencies to help people understand and use design in an effective way in this space. Leon is focused on how design is taken to market. In contrast, Shane is responsible for the internal experience and delivery of design.

Design maturity

Deloitte Australia has yet to realise its ambition to become truly design-led, but it has certainly made huge strides towards it. They have shown how design, when supported, resourced and used to support change, can have a significant impact on an organisation. Shane Currey explains Deloitte's design journey based on three maturity horizons. The first horizon is using design to optimise what an organisation already does and as such, is using it as a tactical tool. The second horizon is using design to develop new products and services. And finally, the third horizon is using design to define its strategy. While in some parts of Deloitte Australia they have reached the third horizon, some other parts are only at the beginning. Nonetheless, it's still a considerable achievement for an organisation with over a hundred and fifty years of heritage dominated by a left-brain mindset. They understand that in order for them to continue to be successful for the next hundred years, they will have to integrate their analytical and rational sides with a more humanistic and insight-led approach.

Deloitte Australia is a remarkable example of how design has broken new ground in a very traditional sector like Assurance and Advisory and shown how it can catalyse a new culture by using design as a *Power Broker* and *Framework Maker*. Over time design has become firmly established as an integral part of Deloitte Australia and is now a key driver of innovation and change.

Notes

1. www.ideasondesign.net/design-for-business/research-conference/design-for-business-research-conference-2015-presenting-papers/design-visualisation-and-storytelling-at-deloitte-and-its-contribution-to-business-strategy/

Bumrungrad Hospital: Transforming patient services

Introduction

Health services are undoubtedly one of the most important and irreplaceable services that we access throughout our lives regardless of country, context or wealth. And yet they are also some of the most complex and often expensive to deliver. This story illustrates how, by adopting a participatory process using simple service design tools, Bumrungrad Hospital in Bangkok has been able to overhaul their service staff training and create new user-led services to ensure they are setting the standards in their sector. They have used design in its roles as a *Cultural Catalyst* and as a *Humaniser* to help them understand and cater to the changing customer needs as well as using design as a *Framework Maker* to guide their service innovation process.

Bumrungrad Hospital, located in the heart of Bangkok, is one of the largest private hospital in South East Asia. It has over 4800 employees, with around 1200 physicians and dentists supported by over 900 nurses. The hospital started out modestly, initially with a 200-bed facility in 1980, replacing it with a larger, 12-storey building, 580-bed inpatient facility in 1997 and finally adding a 300 room outpatient clinic facility in 2008. They treat over a million patients a year and over half are international patients from 190 countries.

The majority of Bumrungrad Hospital's patients are Thai but they have a substantial clientele coming from outside the country, seeking first-class but affordable healthcare. The global medical tourism market is worth up to $55 billion[1] and is growing as much as 25 per cent annually as more people from the affluent west are looking to access world-class healthcare at half the cost. South East Asia is currently one of the leading regions for medical tourism, with many world-class and accredited medical facilities located in Singapore, Thailand and Malaysia. Unsurprisingly, it is a very competitive market and quality of services has become a key differentiator in the market for Bumrungrad Hospital.

Who we spoke to
Varanya Seupsuk, Chief Administrative Officer

Why change?

Although Bumrungrad Hospital in Bangkok is one of the leading private hospitals in Thailand and South East Asia, it needed to find ways to identify, understand and deliver quality service that caters for the changing needs and expectations of a range of different clients. Additionally, they were finding it challenging to motivate and train staff to deliver customer-centric service because the training materials were prescriptive and didn't sit well with the needs of millennial generation staff, who are often entrepreneurial and ambitious. They turned to service design methods and approach to help their staff to understand these changing needs and at the same time to empower them to develop, test and deliver new services based on directly observed needs.

Design roles that enabled change in Bumrungrad Hospital

Types of changes achieved through design

Since 2014

Changing products & services

Changing organisation

What has a design-driven approach brought to Bumrungrad Hospital?

- A co-design approach helps build shared ownership of the idea and acts as a motivator for the entire team.
- The design tools are easy to use and adaptable for different contexts.
- The focus on the customer helps teams stay focused on the human value of their service.

Key challenges

It's in this backdrop that we spoke to Varanya Seupsuk, the Chief Administrative Officer at Bumrungrad Hospital on the use of service design processes and tools to drive service improvements. As Chief Administrative Officer, she has many responsibilities, mainly around managing the support functions in the hospital. However in her opinion, her most important role is looking after customers' experiences. Varanya spoke passionately about how important it is to maintain the high level of service but also how they have to exceed customers' expectations. At the same time she has to ensure that the service quality is sustainable from a business point of view.

'We already provide excellent service but we expect more. This is because our aim is to exceed customer expectations. It's also not just about making services run better, but considering how to make the services more sustainable.'

One of the challenges faced by the hospital is to find ways to equip staff with the appropriate tools and methods to understand customers' changing needs and to find ways to meet them. This agility and awareness of customer needs is important since the growing medical tourism market means that the hospital receives patients from around the world, with differing needs, expectations and cultures. They are finding it increasingly difficult to anticipate needs and respond to them in a timely, patient-centric way.

'Right now every organisation is asking their employees to help them think. So the way of learning, the way of growing your people is different.'

Introduction to service design

'After participating in the Service Design workshop run by TCDC, I became convinced that the tools and methods used in service design will be useful for the hospital to help us understand our customer. For example, it was obvious to me that by mapping out the customer journey with the different touch-points will help us identify areas of improvement and opportunities.'

Varanya and her team were first introduced to service design through a 5-day workshop conducted by the Thailand Creative & Design Center (TCDC). TCDC was established in 2004 by the Thai government to connect and promote the interaction of creative organisations with businesses and government organisations in order to create quality products and services that meet the global market demand. Service design has been gaining traction in Thailand and organisations such as the TCDC has been doing a lot of work in encouraging its uptake by bringing in external service design practitioners to

Stages of transformation

1. Introduction to service design approach and tools by external consultants.
2. Setting up the 'Patient Experience Team' in Bumrungrad to use service design to improve patient experience.
3. Pilot testing the approach with the Heart Centre service team.
4. Scaling up and working with different centres.
5. Setting up a Service Design Library as a repository for tools and ideas generated.
6. Training service design practitioners and leaders through the Bumrungrad Academy.

What can we learn from Bumrungrad Hospital's story?

Co-design processes suited the millennial staff members for whom a sense of idea ownership is key.

Importance of gaining support not only from the executive team but also other key business teams.

Helping teams innovate is part of a larger patient-focused transformative project.

share their work and to run workshops. TCDC has been organising events to introduce and promote service design since 2013 and ran their first the Global Service Design Jam in Chiang Mai, Thailand's second largest northern city, in 2014.

Varanya and her colleagues from Bumrungrad International Hospital and Bumrungrad Academy (set up to train service staff) were introduced to a number of well-known service design tools such as the stakeholder map, personas and customer journey map. They immediately recognised the potential power of these tools to help staff better understand user needs. They also liked the participatory nature of the tools and found that this approach really resonates with their younger millennial service staff. They hypothesised that staff are more likely to participate if they have been involved in its initial conception. This is an example of using design as a *Humaniser*, but in this case, it was not just to connect staff to users but also to connect staff to ideas that they can then take forward. This new strategy of encouraging uptake and ownership of the service by the staff delivering the service has proven to be one of the key factors in ensuring continued engagement from staff throughout the service redevelopment stage.

'One of the key differences between now and before is that a majority of our new staff comes from the millennial generation. I found that with this generation, it's in their nature to constantly challenge and question authority. So we used service design to involve them in coming up with ideas, prototyping the ideas and eventually delivering the service.'

Setting up the Patient Experience Team

'I just felt that the whole process of service design is the answer. So we agreed with my team that we would set up a department using service design to improve the patient's experience. We ended up calling it the Patient Experience Team.'

Enthused by her experience of service design through the workshop, Varanya approached the hospital's executive team to request support in using service design to improve and design new services. She was relying on design as a *Framework Maker*, to give her confidence that it could achieve her goals and ambitions, despite the fact that the precise nature of the outcome was not known at the beginning of the process.

The service design project was also part of a larger service improvement plan that was about to take place across the hospital. As a result, Varanya created the Patient Experience team as a platform to train staff in the use of service design tools and as a place to try out service prototypes. It's a small team and comprises fourteen staff members, mostly involved as trainers. Having gained the support of the executive team, she had the foresight to include the Business Process Improvement (BPI) team to support her work. The BPI team's role is to implement the lean processes for all functions of the hospital and they are involved in putting in place any processes required to deliver a service. Her strategy was not only to get the BPI team onside, but also involved in the process so they begin to understand customer needs and requirements.

Another important aspect was to put in place some key performance indicators (KPI) that will help the Patient Experience (PE) team evaluate how effective their work has been. This is important not only to help the team understand what has worked and what hasn't, but crucially, as a reporting and substantiating tool for the executive team. Some of the KPI's agreed upon were: to track the instances of customer feedback related to service issues and to track the customer satisfaction score. The model would be for the Patient Experience team to introduce service design processes and tools to the different specialist centres through workshops and help them develop a range of service prototypes that could be tested within a 2 month cycle.

The process of service innovation

Armed with support from the executive team and the Business Process Improvement team, their first pilot project involved the Heart Centre, one of Bumrungrad many specialist centres. The first step was to build the team from the Heart Centre that would be involved in the service innovation, and it included the centre's manager and a range of staff at different levels and with different roles in the centre. The plan was to condense the 5-day workshop delivered by the TCDC into two shorter but more focused half-day workshops. TCDC supported the PE team in the facilitation of the workshop.

'We supervised the team for about two months after the initial workshop working with them to develop and test the prototype. Take for example the snack bar. We helped find a food vendor and then supervised them for two months during the prototyping period. At the end of the two months we asked if they wanted to continue with the service. In this example, they wanted to continue, and so at that point we handed it over to the Heart Centre team to take ownership.'

The Heart Centre team's first task was to map out a customer journey of the centre's range of users and to identify the touch points and potential pain points in each of the those journeys. They also created personas to help staff understand the differences in customers' needs and expectations. The customer journeys were then used to help the team generate service concepts that could be prototyped. At the end of the workshop, the team decided which ideas to take forward and trial for 2-months. They also had to decide how they would evaluate the effectiveness of their new service. The team was then encouraged to try out the new service, to track its impact and to then decide (with support from the PE team) if any of the new services or service improvements would be implemented over the longer term. Throughout this participatory process, Varanya stressed that the decisions were made and driven by the Heart Centre team. The Patient Experience team's role was to train, facilitate and enable the team over the two-month period. This bottom-up approach is a key benefit of using a service design approach for the team.

Prototyping ideas

Using the customer journeys enabled the team to identify gaps in the service and opportunities to improve it. For example, they identified the fact that a customer generally follows a set path through the Heart Centre. They will start by registering their visit, then go on to see a physician or be scanned, depending on their condition, before they make a payment and leave the centre. The team recognised that the time taken to move from one stage to another can differ depending on the availability of the physician or the specialised machine. Having identified this problem, the team suggested several ideas to tackle this problem. An initial idea involved using a QR code tag that would be given to the customer when they first registered and would track the patient's progress as he/she moved through the stages. Staff would be able to have an overview of where patients were in the process, and then personally manage expectations of individuals at various points. They could also use the system to spot potential bottlenecks and to find ways to clear them before it became a problem. Another simple idea, which has now been implemented in the Heart Centre, was to play educational videos in the waiting rooms. This would not only offer more information to the patients about their conditions but also help alleviate boredom. This slight change has improved the customer's experience of waiting and the Patient Experience team is now looking to expand the service to other centres and are trying to commission more videos on specific medical conditions.

RESEARCH   DESIGN   TEST   DEVELOP

What are the conditions for impact in Bumrungrad Hospital?

- Gaining support from the executive team
- Working closely with the Business Process Improvement team to ensure they understand the customer needs and requirements.
- Setting and measuring Key Performance Indicators to track outcomes.
- Ensuring the new approach works with existing systems and processes.
- Adapting and simplifying tools to suit Bumrungrad's context.
- Using design to support strategic goals–helping teams create better services is part of a larger patient-focused strategy.

What have been the challenges so far?

- Scaling and speeding up the process of service innovation to the other seventy centres in the hospital.
- Cultivating a culture of innovation and being customer focused.

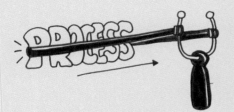

What type of change still needs to be achieved?

- Embedding service design training in Bumrungrad Academy.
- Training more staff to use service design methods.

The Heart team also explored ideas around personalisation of services and finding ways to quickly understand the preferences of high-value customers. Although this was a feature already present in the current customer database, it was not widely used since the staff did not initially recognise its need. However by developing personas that specifically deal with the requirements and expectations of a high-value customer, the team understood how important it was to be able to recognise returning customers and to have their prior preferences logged and easily accessible to staff servicing them.

Another really interesting idea that has emerged from the workshop, with wider implications beyond the Heart Centre, is the volunteer porters' rewards scheme. The hospital has a porter team that provides wheelchairs for customers who are unable to walk. When a customer arrives requiring a wheelchair, the counter staff has to call the porter, who may be busy at the time and as a result the customer may have to wait for the wheelchair arrives. The team suggested a reward scheme for any free staff member who wants to 'volunteer' to fetch the wheelchair for that customer. This idea proved so successful and popular that has since been implemented at the Heart Centre for the last 6 months and the Patient Experience team is looking to implement this scheme across the hospital.

The prototyping phase enabled the team to quickly test if a service is viable or not. One idea that was eventually rejected was the snack bar idea.

'We had to stop the snack bar service at the heart centre, after four months. We were getting patients from the other clinics using the bar and as a result it's not an additional service anymore. No matter how much we were adding more food it was never enough. Instead of being a good service we were getting feedback that we were not providing enough food. So from being a positive add-on, it became a negative experience. So the Heart Centre team decided to stop it.'

Scaling up

'The hospital has more than seventy centres so, based on what we have been doing, it will take a long time to complete the process with all seventy centres. But we can help speed it up by having a service design library because as we work with more centres, we have more ideas and prototypes, which can be implemented directly in other departments.'

The Patient Experience team has now worked with a number of different centres, using a similar model of working. However, despite the success they have been having, they recognise their limitations. As they are a small team, they are only able to work with one centre at a time, and even though the 2-month cycle is fairly short, it would still take another year or more to run the same process with the other 31 centres operating in the hospital. To help speed up this process of service transformation, the team has decided to create a Service Design Library—essentially a repository of the service design process, methods and tools used by the team as well as a repository of service prototypes that can be used in other centres to respond to similar customer needs. So instead of reinventing the wheel twice and three times, the team is able to direct the centre's staff to the library and to see if any of the ideas resonate with their customer journey maps and personas. It also gives them a chance to improve on a previous idea. For example, the volunteer wheelchair reward scheme has been developed into a different solution. The Emergency Wheelchair idea came from the Women's Centre and requires the porters to park two or three wheelchairs at specific locations for emergency use, rather than trying to store them centrally. This idea is now being trialled and illustrates how other teams are encouraged to consider and improve on ideas in an iterative manner.

Value of service design

It is evident speaking to Varanya and hearing her stories that the service design approach is really valued by not just the Patient Experience team, but also by the hospital's executives and the other key business functions. From a management point of view, Varanya finds the persona and the customer journey map to be particularly useful in helping her understand the range of customers expectations and needs, beyond the generic customer survey and questionnaires. She sees it as a perfect counterpoint to the lean process, which is simply focused on cost improvement. The service design approach has offered them a user-focused approach that is fun, easy to understand and to use. The iterative prototyping cycles have enabled them to try out new ideas that can be quickly evaluated and implemented if found successful. They have found a way to be nimble and agile in a large organisation that is service focused. The co-creation nature of service design has also helped achieve buy-in from the staff in delivering new services. The staff also feels enabled to come forward with ideas that will be taken seriously and trialled.

'The business improvement viewpoint was just focused on following a lean process, about cost improvement and improving productivity. But the view is changing now. Service design has been very useful because it helps us understand what the customer wants. The customer journey is very important.'

Aligning design with organisational goals

Varanya recognises that any new ideas or approaches that she introduces can't exist in a vacuum. She astutely identified that the only way for her to achieve buy-in from the hospital executives and the Business Process Improvement team is to be flexible in how it is being used and how it is integrated to the existing system of the organisation.

Service design is not only confined to helping centres improve and offer new services, it's being taught at the Bumrungrad Academy, where service staff is trained. They are teaching staff to use tools like persona and customer journey maps, and creative prototyping methods that will equip staff with the tools required to innovate their services without necessarily needing support from the Patient Experience team. The academy is also developing a young leadership programme at the hospital and part of the syllabus will include service design. Introducing service design tools and methods at the beginning of a staff's Bumrungrad work experience ensures that the user-centric and co-creation approach is embedded from the start. At the very least, it increases the literacy level and appreciation for a more creative, design-led approach in the organisation.

'We're using service design in various places and with different teams. The Bumrungrad Academy themselves are using service design practice, concept and tools to teach our staff. We're developing a young leadership programme at the hospital and the course is really about service design.'

And finally, one of the most effective ways to help people understand what a service design/design-led approach is, is to get them to experience it for themselves. It was therefore important to include members from the Business Process Improvement team as part of the work the PE team did with the different centres. Their participation was crucial, as they were able to offer advice to teams prototyping the services on the most appropriate processes to use. Their involvement also meant that instead of only focusing on the mechanistic process of a service workflow, they were exposed to the human side of the equation through the personas and customer journey maps.

'The Business Process Improvement team normally work on workflow. So, when we brought this idea in, it actually made them agree that the workflow is the customer journey. It's a nice blend.'

The Bumrungrad Hospital's story offers us a glimpse into the beginnings of the design transformation journey in an organisation. It's very inspiring to hear how simple service design tools are being used to transform the way the hospital is managing its relationship with their customers. What is more exciting is that the transformation is not only happening at the service interactions level, but has the potential to transform their systems and embed a culture of innovation. Whether it does remains to be seen, but it seems to have all the required conditions in place for a sustained and impactful transformational journey.

Notes

1. According to research from Patients Beyond Borders.

Spotify: Personalising music experiences globally and moving beyond the start-up phase

Revolutionising the way we listen to music

Spotify is an example of how design has been used as a *Framework Maker*, *Cultural Catalyst* and *Humaniser* to help an already successful start-up become a market-leading global organisation transforming how we listen to music.

Spotify provides access to over 30 million songs and in September 2016 they reached 40 million paying subscribers, double that of Apple Music's base. They offer a free and a paid streaming service and have over 100 million active users on their platform.

Spotify describes themselves as a music company at heart and a tech company by design. Martin Lorentzon and Daniel Ek in Stockholm founded Spotify in 2006. At that time, illegal downloading was at its peak but Martin and Daniel's vision was to make music available to everyone in way that was legal and fair both to customers and to artists. In the early days, the focus was on building the technological capabilities and negotiating licensing agreements with the music labels. So the first Spotify app, launched in 2008, was mainly a music library with a search box.

Spotify's subscriber numbers has been steadily growing since 2011, from a million subscribers to over 40 million in 2016. Their pace of growth

Who we spoke to
Sofie Lindblom, Global Innovation Manager
Ste Everington, Senior Product Designer

Why change?

Spotify started life as tech start-up in 2008 and has been growing rapidly in the last few years as the music streaming industry matures. The rapid growth of the company has left design under-resourced and less prioritised. At the same time they needed a more structured approach to innovation and product development in order to leverage their latent creative culture while still continuing to be a market leader.

Design roles that enabled change in Spotify

Types of changes achieved through design

Since 2012

Changing products & services

Changing organisation

What has a design-driven approach brought to Spotify?

- A more structured and proven innovation strategy.
- The innovation process has also helped them work more cross-functionally through different teams and disciplines.
- A more considered approach to their product development.
- A more user-focused approach.

in the last two years has been significant, for example increasing 5 million subscribers within the space of 6 months. This accelerated growth has been due to a number of reasons. The concept of streaming is becoming more widely adopted. Unlike 10 years ago when Spotify first started, the millennial generation is more likely to accept the model of streaming rather than owning music. Spotify has also expanded to many other countries globally and that has helped to increase their market reach.

Spotify has recently re-positioned their products from a music streaming service to a lifestyle product and based on their 'moments' strategy. The goal is for music to play a bigger role in a person's life at different moments of the day. For example they have recently launched two new product functions in their app, Spotify Running and Spotify Party targeted at two specific use cases. Spotify Running will match the music to the beat of your feet as you run. It helps the user keep tempo and provides relevant music based on the pace of running. Spotify Party allows collaboration and mood setting especially in a party context as well as intelligent cross-fading between tracks for a more 'DJ' like experience.

Design in Spotify

The story of how design has been used to transform Spotify can be discussed in a number of ways. The first is at a more strategic and global level, in how Spotify has tried to re-establish design as a core function alongside their tech and product functions. Secondly, design has been instrumental in helping Spotify develop and drive their innovation strategy. Design has acted as a *Cultural Catalyst* and *Framework Maker* in helping Spotify build an innovation culture. Design has also acted as a *Humaniser* in helping product teams in Spotify to focus on the users.

Design has always been present in Spotify, but up until a few years ago it has been under resourced. There were a number of reasons for this. In the early days, focus was on the technical feasibility of the app as well as negotiating licences with music rights holders. From that point onwards, it was a case of building the market and developing a viable business model while concurrently improving the product. Design hadn't grown at the same scale as the rest of the company. So while Spotify was known as a disruptive force in the music industry, their products in the early days were not particularly known for their usability and user experience. Design was often relegated to finishing ideas rather than being part of the conversation around the strategic direction of the company. It wasn't until 2012 that design started to play a more prominent role in Spotify with the appointment of Rochelle King as their VP of Design and User Experience.

Stages of transformation

1. Rochele King's appointment in 2012 brought design to the fore and rebalanced design capabilities with the other core tech and product functions of Spotify.

2. Major overhaul of Spotify's mobile and desktop apps and the launch of Spotify's design principles in 2014.

3. Conversations started in 2015 around innovation practices in Spotify led to the realisation that they needed a more structured and holistic way to manage innovation.

4. As a result, the Innovation Team was set up in 2015 and Sofie Lindblom was appointed Global Innovation Manager.

5. The Data, Insights and Design organisation in Spotify was created at the end of 2015. It was aimed at bringing together areas focused on the company with areas focused on product experience.

What can we learn from Spotify's story?

- It is important to have an exchange of expertise and approaches in order to maximise the potential of design's impact.
- Embracing failure as a natural element of the innovation process.
- Design helps with establishing innovation structures as a company grows beyond its start-up phase.
- Innovation can be owned by everyone.
- It is important that design is aligned with the organisational culture, structure and goals so it delivers value immediately.

Realigning design

When Rochelle first started to build a design practice in Spotify, her main goal was to align design with the larger and more dominant tech and product aspects of the organisation. For her, it's fundamentally about getting everyone to speak the same language and to explicitly align around the same goals. It was important to build empathy with each other. However, it's not just a one-way alignment where designers have to understand the product and tech perspectives, it's also about helping their partners learn how design works to create a more holistic user experience. At the same time it was also important to establish some clear design principles that can be used to communicate what design is doing as well as to drive its direction in the company in line with the overall company strategy. They came up with six design principles that became the backbone of company's rebrand and the redesign of their app in 2014. These principle are: Content first, Be alive, Get familiar, Do less, Stay authentic and 'Lagom' – a Swedish word to describe a perfect balance between not too much and not too little. These design principles still epitomise Spotify's product philosophy. The redesign was extremely well received by customers and made an explicit statement that Spotify was becoming more design savvy. It changed the way design was seen internally in Spotify and it let the designers focus on solving bigger design problems that are beyond visual styling.

Building up design capability

Design was severely under-resourced when Rochelle joined Spotify. In theory, project teams (or squads as they are called) in Spotify always comprised a tech owner, a design owner and a product owner. However in reality not every squad would have a designer since there weren't enough designers to go around. As a result, designers were thinly spread across the teams and it was difficult for them to grow and have a strong enough voice to shape the company's strategic vision. Realising this, Spotify started to recruit more designers, especially hiring people with user research and prototyping skills. Rochelle also established a user research team, which works across the squads and offers its support when needed.

Giving a voice to the user

One of the key differences design has made to Spotify and to how they develop products is by its focus on users and a quest for insights linked to the intricate and often hidden needs of users. Ste Everington, a senior product designer, reflects on how the attitude towards users and user testing has changed in the last few years at Spotify.

'When I joined Spotify, the concept of testing a product before shipping was unheard of. Instead, a product would ship with nothing more than a gut feeling for success. Now that's completely changed. Today, all teams understand the value of user research and it's tightly integrated into our design/development process.'
Ste Everington, Senior Product Designer

It was also important that the tech and product team in the squad were also able to empathise with the user. This in turn helps them to empathise with how designers think and work. This condition allows for the entire team to focus on the user and work collaboratively to solve problems.

Framing innovation in Spotify

Although Spotify currently has over 2000 employees globally and is considered a 'mature' corporate organisation, they are in a sense still operating like a start-up where ideas can come from anyone and anywhere and can be developed independently. While this creative culture has been key to the early success of the company, it has also been a challenge to bring together different products and ideas into a coherent strategy. Although people were engaging with innovative practices, they weren't exactly aware of it nor were their activities targeted or aligned. At the same time, the founders were also concerned that they would lose their creative culture as the company expanded. They wanted to ensure that Spotify's culture was not only protected, but also developed in order to cope with the growth of the company. To achieve this perfect balance, a new innovation team at Spotify was created to help structure innovation in Spotify and make it a core part of their mission and vision.

The Innovation Team was created in Spring 2015 and Sofie Lindblom was tasked to head the team. Sofie's masters thesis focused on how Spotify worked with innovation and it helped spark the initial conversations with Daniel Ek and other Spotify teams about how innovation happens in Spotify. Innovation projects from across the organisation were gathered to understand what they were working on globally and to get a holistic overview. They conducted qualitative and quantitative research to identify the barriers to innovation, how they define success for innovation, and to gather information about what work streams to focus on.

INNOVATION
— TEAM —

A framework for innovation

Innovation in Spotify is structured through four work streams: 'Culture of Innovation', 'Strategic Innovation', 'Data-driven Innovation' and the 'Think It Framework'. The first stream is related to creating a culture of innovation, which in itself is a huge challenge. Spotify recognises the importance of supporting a culture of innovation because without it, practices will not happen.

'This is the one I struggled the most with because culture can be such a fluffy thing and it's hard to measure if you're making an impact or if the things you're doing are successful.'
Sofie Lindblom, Global Innovation Manager

They have two key ways to build innovation practices. Firstly by communicating all the great things they are already doing to inspire and motivate others in Spotify. Secondly to engage everyone in activities that foster creativity and create forums for innovation to happen. The benefit of Spotify coming from a start-up background is that they already have pockets of innovation happening throughout the company. For example, they have been running Hack Weeks where everyone is given a week off to work on something they are passionate about. This has been very successful and in the past many new ideas and products have been born during these weeks. They are building on activities like these and providing further support in the form of sharing articles, talks and books about innovation.

The 'Strategic Innovation' work stream focuses on looking at what will happen with the music and the tech industry in three, five and ten years from now, and how they can influence its direction. The aim is to identify which micro and macro trends to focus on, what technology to invest in and who their future customers would be. It's about stepping back from the day-to-day operations and casting their net into the future.

INNOVATION IN SPOTIFY

CULTURE OF INNOVATION

STRATEGIC INNOVATION

DATA DRIVEN INNOVATION

THINK IT FRAMEWORK

THE
THINK IT
FRAMEWORK

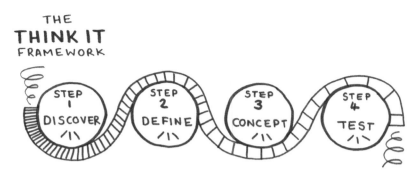

STEP 1 — DISCOVER

STEP 2 — DEFINE

STEP 3 — CONCEPT

STEP 4 — TEST

The third stream is 'Data-driven Innovation'. Sofie's Innovation team are not directly involved but are working as a bridge between their analytics and implementation teams. It's about identifying what they know in order to make knowledge available and subsequently being able to pull the right teams together.

The final work stream is the 'Think It Framework', which is essentially Spotify's innovation process. Design's influence has been most keenly felt in this area. The 'Think It Framework' has been developed with a design thinking approach and designed to help teams navigate more effectively through the early generative phases of a project. It is part of Spotify's product development process that consists of 'Think It, Build It, Ship It and Tweak It'. The majority of Sofie's work has been to facilitate the use of the 'Think It Framework' with different teams. The 'Think It Framework' consists of four steps: 'Discover It, Define It, Concept it and Test It'. Each step has its own set of tools to be used, depending on the type of problem. Many of these tools are recognisable and commonly used tools in a design process such as personas, customer journeys and storyboarding while some tools have been developed specifically for Spotify's use such as 'How might we' cards, Concept cards and Pitch template.

'I worked very closely with a designer in London to build the Think It Framework, the different tools in it and to upscale it in the early stage. The framework is based on design thinking. The interesting thing is that designers are already very familiar with this way of working. But for other parts of the organisation it's not necessarily true that they have ever worked in this way or knew of this way of working. So I think in terms of transforming the organisation with design, I think that's where it's been having the most impact. It's teaching people a design and creative thinking framework.'
Sofie Linblom

Sofie's team acts a centralised resource working like an internal consultant that can be called on by any team. This model is helpful for a company that is globally distributed and it also means that they can help prevent potential

overlaps and help match-make different teams depending on the expertise and knowledge required.

What worked

There are a number of factors that enabled design to be adopted successfully at Spotify. Coming from a start-up culture they believe that ideas can come from anyone and anywhere. Creativity is not just confined to the design or the tech team; everyone can and should be creative. This open view of creativity and innovation has meant that the innovation process and product development is a much more collaborative effort. Innovation isn't just owned by a specific team or discipline, it's a joint responsibility. The culture is also really focused on learning. They don't just want to launch a product that's good; they want to learn why it's good and why users are using it in that way.

The start-up mentality has also influenced how Spotify manages risk. They are not afraid to halt a product under development, if they find that it's not working. For many companies this would be very difficult to do, since a lot of time and money have already been invested in the development process.

'At Spotify, you try a lot of things and if they fail there's no point in trying and trying, and hoping that it succeeds. So we're not really afraid to pull a product, if it does not work. I think it's very difficult to be able to do that as a company, especially if you've invested a lot of time into something. I really value that Spotify is not afraid to do that.'
Ste Everington

Collaborating with other core functions

In November 2015, a new organisation called Data, Insights, and Design was set up in Spotify as an experiment to combine design, data and user insights into a single department. It's made up of a few different teams. Product design is focused on defining the user experience. The Product Insights team includes user researchers and analysts that are focused on understanding how their products are performing by gathering and analysing big and small data. The Analytics team is focused on studying analytics across the entire company and to make the company data-first by encouraging best practices, culture and

What are the conditions for impact in Spotify?

- Support from the executive team (buy-in and part of the corporate strategy).
- Rebalancing design as one of the core functions of a product team.
- Better resourcing of design.
- Making a collaborative effort in adopting design-led approaches, rather than ring-fencing creative activities.

What have been the challenges so far?

- Growth of the organisation.
- Bringing together different practices.
- Creating a platform and common approach without losing diversity and autonomy.

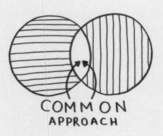

What type of change still needs to be achieved?

- Establishing shared and consistent innovation and design practices.
- Creating and maintaining a culture of innovation.

tools. Sofie's Innovation Team is also part of this new setup. This new organisation shows that design is no longer considered as an afterthought and instead is part of the backbone in shaping Spotify's future products, strategy and culture.

Moving forward

Spotify has been through a period of rapid growth in the last 3 years. They have become more design savvy and have looked to leverage the role of design as a strategic tool. Design has helped them be more thoughtful in how they execute projects and what they spend their time on. They are more confident in the things they are investing in because they now have a process and structure they believe will give them the most effective results. Design capabilities have been strengthened and integrated better with the tech and product team. They have done this not only by using design as a *Framework Maker* to guide their innovation process but also by using design as a *Cultural Catalyst* to rally everyone in the squads to listen to the user. Design has given users a voice in Spotify and acted as a *Humaniser*. The customer's voice is now a constant in the development process and has helped shape more innovative products from Spotify. Design has not only been aligned with Spotify's business goals, it is well placed to drive it.

Itaú Bank: Learning, building and dreaming a new innovative culture

Introduction

Itaú Bank is a long-established bank in Brazil with over 5,000 full-service branches and 28,000 ATMs. It is the largest bank in Latin America and is amongst the 30 largest banks in the world based on market value. It is the leading, privately-owned bank in Latin America and has approximately 100,000 employees, 140 million clients and operations in 20 countries throughout the Americas, Asia and Europe. They are a universal bank with a range of services and products serving a varied client profile: both individuals and companies of all sizes, from major transnational groups to local micro-entrepreneurs.

Similar to other sectors that we have touched on in the book, the banking sector is ripe for disruption and change. This case study illustrates the importance of having an explicit and clear innovation strategy, especially important in a sector that is conservative and adverse to risk. By using design as a *Framework Maker*, the organisation is able to embark on this process with confidence despite not being able to predict the outcome at the start. Following a framework is only the start, since the overarching aim is to catalyse a more innovative culture in the bank. Itaú was not only using design as a *Cultural Catalyst* for its own organisation but is aiming to have an impact on the banking sector in Latin America through a broader innovation agenda to build partnerships with different industries and companies.

Who we spoke to
Ellen Kiss Meyerfreund, Director, WMS Innovation Team

Why change?

The merger between Itaú and Unibanco in 2009 raised three questions: how does the bank retain its market leader status in Brazil, how does it continue to stand out in an increasingly complex and regulatory-driven market place, how can it continue to develop ever-higher levels of financial services products and services while still increasing the levels of client loyalty? The answer–to create and follow an innovation process and at the same time put in place a strategy, structures and resources to support the growth of an innovative culture.

Design roles that enabled change in Itaú

Types of changes achieved through design

Since 2009

Changing products & services

Changing organisation

What has a design-driven approach brought to Itaú?

- Identify new opportunities and develop them into market ready products and services.
- Help silos work together better.
- More effective anticipation of future challenges and customer needs.

An established brand known for its innovation since the beginning

Itaú Bank began in 1945 and has a strong history of innovative technology use. It partnered with a technology company, Itaútec to create the first ATM machine in South America. Having engineers as the founders helped establish a strong engineering mindset in the bank's DNA.

The push for a more user-centred innovation focus came from the bank's 2009 merger with Unibanco. The merger made Itaú the largest financial conglomerate in the Southern Hemisphere and the 10th largest bank in the world by market value. As well as the broad challenges of such a merger, on the private banking and asset management side, further challenges were identified. The bank's leadership posed these challenges: how does the bank retain its market leader status in Brazil, how does it continue to stand out in an increasingly complex and regulatory-driven market place, how can it continue to develop ever-higher levels of financial services products and services, and how can it increase levels of client loyalty? The answer was to focus on and make innovation a culture and mindset across the bank. As a result, the Wealth Management and Service (WMS) team started to focus a significant amount of its energy and resources on innovation as a central part of its business planning and thinking.

Why innovate?

The main trigger for innovation in Itaú is the realisation that the knowledge, experience and awareness that have brought them this far may not necessarily prepare them for the future. They have a saying: 'What brought us here will not take us there'. It is this awareness and foresight that has been the driving force for the bank to continually innovate.

There are a number of significant current and future challenges in the banking industry. Ellen Kiss Meyerfreund, Head of Innovation at Itaú Bank pointed out three key challenges faced by the bank. Firstly, the banking sector is highly regulated. Although this offers some protection to existing banks from new competitors due to the high barrier to entry, it also makes it very difficult to innovate. Let's take the example of creating an online account, which we now take for granted. Itaú wanted to offer their clients the ability to open and manage their account online. Ideally the set-up should be entirely online but, due to regulation requirements, a physical signature is required to complete the process. Apart from regulations, there are also cultural and structural legacies that established banks have to deal with. In contrast, start-ups have a clean slate to start with and are able to develop specific solutions for niche problems. PayPal is a good example of this. They did not exist fifteen years ago but have now completely dominated the online payment market. New players entering into the market has resulted in the banking value chain becoming more and more fragmented.

Stages of transformation

1. The Wealth Management and Services team (part of the private banking area) set up its own dedicated innovation team in 2010 to drive innovation from within.

2. IDEO introduced design thinking into Itaú and helped them structure their strategy into three stages of 'Learn', 'Build' and 'Dream'. The Wealth Management and Services (WMS) Innovation Team expanded on these stages and developed the 'Identify' 'Develop' and 'Implement' pillars.

3. Implementing the strategy and activities within the three pillars.

What can we learn from Itaú's story?

Itaú is driven by an explicit innovation strategy and process adapted to a Brazilian mindset.

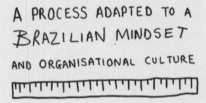

The change process is supported and facilitated by a programme of activities, events, training and resources open to all staff.

Itaú has been very visible in terms of its ambition to be an innovative organisation internally as well as externally to ensure it becomes part of the company's narrative.

It is also increasingly difficult to manage customer expectations in an era where customers are able to 'facebook' and' tweet' directly to companies. Increasingly Itaú's competitors are not other banks, but other digital service companies. It has become a huge challenge for banks to cope with the increasing level of customer expectations and to deliver digitally enhanced experiences at the level that customers now expect.

There is also a challenge that is specific to Latin America. The social and economic gap between customers in Latin American countries is still very wide. This makes it all the more challenging for Itaú since it offers private as well as retail banking services. They want to deliver high-level experiences for clients at the top of the pyramid but at the same time appeal to customers at the lower end of the pyramid who perhaps lack financial education and have a mistrust of banks. So the solutions and products offered by Itaú have to be very broad in order to meet different customers' perspectives and expectations.

The Innovation Team at Itaú

To help Itaú face and address these challenges, the Wealth Management and Services team (part of the private banking area) created its own dedicated innovation team in 2010, starting with a team of fifteen staff members recruited from a number of sectors outside the banking industry. Part of the early set-up process also included orientation and learning trips to sites of innovation such as Silicon Valley. Since the beginning, the innovation team has worked with two significant principles in mind: to change mindsets within the bank by challenging conventional bank-type thinking and to involve clients as much as possible in the process.

Ellen Kiss Meyerfreund took over as the head of the Innovation Team at the Wealth Management and Services (WMS) area in 2014, having previously worked as an Innovation and User Experience manager in Itaú. The WMS Innovation team's role is to work as a partner with the bank's internal business units in order to create products and services that are more client/user-oriented and more importantly to bring more value to the business. Her team currently has grown to seventeen people and they are generally grouped into three areas. The first group are people who are more business orientated and have an understanding of business. The second group are mostly made up of psychologists, social scientists and design researchers who understand people and are good facilitators. And the final group are mainly graphic, visual and

UX designers who are very good at turning insights into design ideas and working on tangible design outputs.

The team works to coordinate and support a range of projects aimed at simplifying processes and enhancing services by proposing alternative solutions. The team prioritises clients' user experiences and uses IDEO's inspired design thinking methodology to drive innovation by making products and services that are desirable, feasible and viable. They begin by designing what is desirable from the clients´ point of view, then they start to work on the solutions' feasibility, finally, once desirability and feasibility are achieved, they move to the final phase where the financial perspective is addressed. In these and other ways the WMS innovation team is an exception to the standard rules of banking and private banking.

When the Innovation Team was set up in 2010, its aim was to catalyse and instil a culture of innovation. They tried to achieve this in two ways, firstly by working on culture-related projects and secondly on projects with a specific work/result focus. Many of the culture-related projects involved events, lectures, training and visits to firms such as FIAT, Google, Coca-Cola and IBM to help staff understand what innovation is and how it has been applied. In contrast, the project-related innovation had specific return on investment principles linked to project outcomes. Without this distinction, it would be harder to identify innovative practices and evidence their impact.

An innovation strategy

The Itaú Innovation strategy was developed with the help of IDEO in 2010 to enable the bank to develop an innovation culture. At that time using design in such a way was considered visionary in the banking sector, especially in Brazil. IDEO introduced design thinking to Itaú and helped them structure their strategy into three stages: 'Learn', 'Build' and 'Dream'. The Wealth Management and Services (WMS) Innovation Team expanded on these stages and developed the 'Identify' 'Develop' and 'Implement' pillars. The WMS Innovation Team went on to develop its own three-step innovation process that consisted of 'Inspiration', 'Ideation' and 'Development'. This process helped provide a practical and easy to follow guide for teams working with the WMS innovation Team.

Design has always been present in the bank, even prior to 2010. However, it was mostly confined to visual design, packaging and branding within the marketing team. Since the introduction of design thinking and a focus on user experience, design has been much more valued and its role has expanded to include the design of experiences (through digital products and services) through to organisational change (through its innovation strategy). Design is being used as an approach to help Itaú's staff to understand customers. It has also helped the business units to work in a more collaborative way amongst themselves as well as with other teams.

The innovation strategy and process provided the WMS Innovation Team and the wider Itaú community with an understandable process and structure for innovation and transformation. It acted as a *Framework Maker* for the team. It gave the organisation confidence in the outcomes and helped teams manage risk. They needed an easy to understand process to help them create an innovative culture from the ground-up. It was also important for the innovation team to not just focus on the tools but to adapt the process to a 'Jeitinho' mindset, which is a typically Brazilian approach borne out of limited resources to accomplish a task by being creative (sometimes circumventing rules) and inventing simpler ways to do things.

'You really need to create ideas that are based on clients, technology and business. I really love that framework from IDEO. It's really important that what we create is desirable for clients, viable as a business idea and technologically possible. You really need all those three elements but it is difficult since people tend to focus on one or two but not all three.'
Ellen Kiss Meyerfreund, Director, WMS Innovation Team

Providing a framework to innovate

The innovation strategy at Itaú consists of three main pillars: Identify, Develop and Implement. The 'Identify' Pillar focused on the learning requirements of the organisation. Building a culture of innovation and shifting the mindset across Itaú meant first beginning with learning what it means to innovate and hence the first pillar is heavily focused on education and training. There are currently three initiatives running under this pillar. Firstly, the Driver 2020 initiative is aimed at keeping Itaú connected with the external world. It helps teams keep abreast of innovative practices and trends happening outside the organisation and sector. It does this by following awards, sending staff to attend events (like the South by Southwest conference) and mapping innovation projects across different sectors around the world. It offers the most direct way of bringing expertise into Itaú.

The second initiative is the creation of an Open Innovation Platform. The platform is a vehicle for the Innovation Team to go beyond the bank and seek collaborations with external partners. This is increasingly important because companies are moving to a more integrated model of service delivery. A person doesn't just interface with one organisation in his or her daily life but instead uses a range of external services from a number of different organisations. So the Open Innovation Platform enables Itaú to collaborate with companies and universities and potentially build integrated services for their customers.

The third initiative under this pillar is the 'Insights' event described by Ellen as an internal 'TED'[1] event. It consists of a series of workshops, held once a year, which focus on one or more themes. At these events, the bank's employees have the opportunity learn at an off-site environment to encourage creative thinking. Insights events have focused on areas such as education, architecture, urbanisation, urban environment, and so on. When it first started, the Insights events were attended by over 100 internal employees. However, it has since expanded to over 1,000 attendees who include employees as well as clients and staff from partners and suppliers.

In addition to these three explicit initiatives, Itaú has been committed to putting employees through a course on innovation run jointly by the Innovation team with IDEO. The Innovation Drops course allows the bank's employees to learn and apply innovation approaches to their daily lives. The innovation team has also been training staff on lean start-up methodologies, a type of thinking not typically applied within a large bank or corporate context.

Developing opportunities into ideas

The second pillar of the Itaú's Innovation Strategy, 'Develop' is focused on working closely with Itaú's internal business units to identify business challenges and innovation opportunities. The main aim is instil a proactive approach that will help teams forecast and anticipate coming challenges. There is often a huge gap between the mechanistic and rational aspects of a business and the more humanistic aspects of understanding people's needs. The Innovation Team's role is to bridge the business and user needs and help teams identify opportunities and support them with the skills to become more innovative in their practices. Innovation projects can take two forms. The first is a large strategic project usually lasting four months using a standard innovation method that consists of three different stages: 'Inspiration', 'Ideation' and 'Development'. At the 'Inspiration' stage, teams are offered techniques to help them frame opportunities, conduct user-centred research and uncover insights on client behaviours and needs. At the 'Ideation' stage, they are offered methods for collaborating, a structure for testing ideas, guidance on concept generation and help with visualising use scenarios. And finally in the 'Development' stage, they help teams prototype and test ideas as well as develop an implementation plan.

Innovation projects can also take another shorter form called 'Innovation Sprints'. These projects are focused on a specific challenge and adopt a shorter 5-day project timeline by using Google's design sprint methodology. Potential stakeholders are mapped on the sprint and a project room is created exclusively for members to work in. This new approach was introduced in 2015 and has been a very interesting learning experience for the team. The team is continually adapting and improving the process but the shorter format allows the team to try out ideas quickly and in a more exploratory manner.

The team also run an annual competition called 'Challenge'. Everyone in Itaú is encouraged to submit ideas that have the potential to improve client experiences and ultimately, the bank's financial results. It's a six-month process with several stages. Initially twenty ideas are selected, developed and pitched. In the second round only ten ideas are taken forward. At this stage they receive additional support and mentoring from the Innovation Team and are assigned a dedicated project room within Itaú's innovation campus, Inovateca. The project team is staffed with a mix of skilled people from the innovation, marketing and technology teams. Each project is required to engage clients as soon as possible. Finally, the idea's owners are given the opportunity to present their ideas to the board of directors. So far four editions of the Challenge project have been completed and this has resulted in over 400 ideas submitted. It's a win-win situation for everyone. The bank is able to have a first look at new ideas while staff gains confidence through learning how to develop, prototype and present ideas. These tactics: creating different formats to help teams turn insights into ideas supported by a proven process, expertise from the WMS Innovation team, and a physical space for collaboration, are key to using design in a *Cultural Catalyst* role.

Build offerings, business and experiences in new ways

The final pillar of Itaú's Innovation Strategy is arguably the most important, since it's linked with implementation. It is pointless to change mindsets and embed an innovative culture if these initiatives do not lead to actual changes in practices. Hence the 'Implementation' Pillar is really key to the success of the strategy. It is about identifying the current and future barriers to implementation and coming up with a plan to overcome them. Ellen recognises how difficult it is to implement new ideas in an organisation. The team is working closely with IT in a more integrated way and have started using AGILE and a LEAN start-up approach to help teams implement digital solutions faster.

Catalysing an innovative culture
Creating an awareness of innovation

How can you use design to catalyse an innovation culture? The Innovation Team at Itaú has used a design thinking inspired framework: Learn, Build and Dream to help them expand and develop their innovation strategies under a number of pillars and initiatives. The initiatives that we mentioned earlier all contribute to supporting, creating and enabling new innovative practices to emerge and become the norm. We learnt that it is very difficult to catalyse a culture without first inspiring others to think and act differently. Itaú has done this in a number of ways. The Innovation team brought in external expertise and used examples from the outside to get people excited about the value and possibilities of an innovative culture. Ellen has likened it to starting a movement.

What are the conditions for impact in Itaú?

- A history of being innovative (especially in technology use).
- Awareness and foresight to prioritise innovation practices.
- Support and leadership from the executive team.
- Having a focused, flexible and explicit innovation strategy.
- Creating resources to support the initiation, development and sustainment of an innovative culture.
- Bringing people from the outside to question existing practices.
- Maintaining an 'outsider' but relevant point of view to the organisation.

What have been the challenges so far?

- Banking is a highly regulated sector.
- Changing customer expectations and the relationship with their banks.
- Value chain is being dispersed due to new niches offered by start-ups that have not had to deal with years of cultural and structural legacy.
- Catering to a broad range of customers in Latin America.

'It's first creating an awareness that innovation is important, that the world is challenging, every industry is going to be disrupted so it's important that we have to innovate. The first step is to learn what innovation is and then teach people how to do it.'

Once you have created this awareness and a desire to innovate, it's then about getting them to think differently. The short five-day sprints aren't so much focused on generating a viable idea as they are a useful mechanism to get people thinking in a much more user-oriented way and to identify ideas that can generate value for the business and for society. Getting them to learn by doing is a really key part of the learning process since it builds up their capabilities and builds confidence in their creative abilities.

'It's sometimes pointless to get them to think differently if they don't feel empowered to see the idea go forward.'

Questioning current practices

The 'Insights' and 'Drive 2020' initiatives are examples of bringing ideas from the outside not only to inspire but also to question existing practices in banking. Innovation is used as a powerful transformational agent since it is seen to be radically different from the approach usually found in the banking sector.

'We have to be constantly provoking, challenging and questioning people all the time–Why did you do it in that way? Is it really what you want? It this really what the client wants?'

Similarly the Challenge competition provides a platform for people to try out new ideas and this often means challenging established practices. The competition allows for a 'safe' space to innovate but also provides an outlet for riskier ideas to develop and flourish. Design's role as a *Friendly Challenger* is to question and challenge in a less confrontational way.

Creating an innovation space

The Innovation Team also recognises the importance of a physical environment to support new innovative and collaborative practices. With this in mind, the team created special project spaces called 'Inovateca' to house innovation projects. These are dedicated spaces, which can be, personalised and changed as needed by the project team. As a result of this effort, other bank sites also created 'Inovateca' spaces with the same goals in order to leverage the innovation atmosphere within their teams. This, once more, is a very different experience from a traditional project environment within a bank. As well as the location, part of this process also includes bringing together staff with different backgrounds and skills on projects. This might include staff from other departments including innovation, marketing, design, technology, private banking, and so on. Having a separate space to collaborate is particularly important since it helps create a neutral space for people from different departments to work together. Collaboration is really key and projecting a level playing field is important in helping maintain an open culture.

Having top-down support

Although the impetus to embrace an innovation culture has to happen at all levels of the organisation, it's vital to have support, direction and leadership from the top. The Innovation Team was fortunate to have the early support of the Vice President in the Private Banking team. He has been instrumental in pushing the innovation agenda at Itaú. Without explicit backing from senior management, progress has been slower in other parts of Itaú.

'It's really hard when you start from the bottom up. We only got to where we are now because we have had the support of our VP from the very beginning.'

Key to success

Although the WMS innovation team started within the private banking side of the bank, it is noticeable and significant to see how its activities have now been broadened to include other units of the bank. In total, the bank has around 300 staff, directly or indirectly working around innovation in areas such as technology, digital channels and marketing. These different units also work together to share and leverage their respective experience, ideas and projects.

At a more personal level, Ellen has seen how the transformation has changed the way people worked in Itaú and also how Itaú is perceived from the outside. She says:

'I am seeing people using the tools that we have introduced even when we have not been involved in their project, for example the use of personas. Their vocabulary is more integrated and we have seen some of the teams running innovative sprints by themselves. I have also been trying to position Itaú as an innovative company with the aim of attracting people who might not necessarily want to work in a bank. It is definitely working. I just recently hired a designer who used to work in Nike. It is amazing that he wanted to leave Nike to work in Itaú.'

There are a number of factors that have been key to the Innovation Team's success at Itaú. Firstly, they needed to find ways to identify new opportunities. Tracking trends and bringing in expertise from the outside (through collaborations with universities and other organisations), prepares Itaú for challenges ahead. Secondly, it was important to link innovation with their business strategy model. This meant working closely with the different business units in Itaú to understand what they do and how they do things. The third factor is to work with these business units on hands-on innovation projects. It helps teams experience using the tools and processes first-hand and, as a result, to learn what innovation is. It was also important to support teams throughout their innovation journey from the start (the concept stage) to the end (the implementation stage). Finally, it is really important to establish success metrics for each project to ensure there is a way to evidence value.

Challenges ahead

Although Itaú's Innovation Team has had a lot of success in building and supporting an innovative culture at Itaú for the past five years, Ellen knows there are still three main challenges ahead. Her immediate goal is to initiate and work on more strategic projects. Although the Innovation Team has been working on many projects, Ellen wants to focus on larger more ambitious projects that will help Itaú be at the forefront of the sector. She wants to focus on longer-term value creation for the bank. The second challenge is to have more impact across different industries by building a unique innovation agenda. She wants Itaú to be a leader across different sectors and build synergies with other organisations. And her final challenge is to continually find ways to challenge and change existing systems and culture in the bank to ensure faster implementation of ideas. Ellen believes that things will really start to change when well-established companies start to work closely with start-ups.

Innovation is recognised as a key asset for Itaú and is considered part of their internal culture. This was evident as early as 2010 when Fast Company voted it one of the top ten innovative financial companies, the only bank to be on the list in that year. Itaú places great importance on creating products and services from a client-focused point of view and is continually thinking of ways of making them more simple, efficient and practical. While it remains a challenge to be innovative in a highly regulated sector, Itaú has illustrated the importance of adopting an innovation strategy that is strongly driven by the philosophy of helping people learn through doing. This is also helped by having a framework and process that is easily understood, implemented and, importantly, that can be adapted to suit the specific culture and mindset of the organisation.

Notes

1. TED stands for Technology, Entertainment and Design is a global set of conferences run by the private non-profit Sapling Foundation, under the slogan "Ideas Worth Spreading". TED was founded in 1984 as a one-off event; the annual conference series began in 1990.

What type of change still needs to be achieved?

- Faster implementation of ideas and quicker to market.
- Focus on longer-term strategic change projects.
- Building an innovation agenda and setting up synergies with different industries and companies in their network.

Telstra: Becoming a customer-centric organisation using design thinking

From an engineering company to a customer-centric service provider

Telstra is Australia's leading telecommunications and information services company with over 36,000 employees. They dominate the Australian market, providing 16.7 million mobile services, 7.3 million fixed-voice services and 3.1 million retail fixed broadband services.

While Telstra has long been the leader in the telecommunications industry in Australia, in 2009 under then new CEO David Thodey, they embarked on a mission to become a leader in terms of customer service as well. Their mantra was to put the customer at the forefront of everything they do. That is a mantra that remains their number-one priority today and we can see the results that it's delivering. Telstra has increased at least 10 per cent of their market share in the mobile business in the last five years (according to GSMA Intelligence) and the public perception of the company has had a dramatic turnaround.

They are proud to say that the Design Practice team has been part of that change and has played a key role in helping design staff achieve their potential as a customer focused organisation.

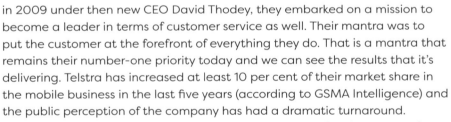

Who we spoke to
Cecilia Hill, General Manager, The Design Practice
Scott Barclay, HR Director, Organisational Change and Transformation Design

Why change?

Telstra wanted to cement their market position as the leader of the telecommunications industry in Australia and future-proof their growth by becoming more customer-centric. They also wanted to encourage and grow an innovative culture in Telstra to keep them at the forefront of market challenges.

Design roles that enabled change in Telstra

Types of changes achieved through design

Since 2013

Changing products & services

Changing organisation

What has a design-driven approach brought to Telstra?

- A method to involve customers.
- Encouraging early and continuous prototyping of ideas.
- A de-risking tool.
- Using customer insights to focus the design goals.

Moving up the Ladder

The role of design in Telstra has been historically linked to usability and user experience. But in 2012 the focus of the role changed with moves from a specialised user interface design team to a team focused on a more strategic application of design. They restructured the team and set about showing the value of design to the organisation. Their staff were given the challenge to complete five projects in twelve months in order to demonstrate how design and design thinking can be used to help teams work more innovatively in Telstra.

For Cecilia Hill the General Manager of The Design Practice at Telstra, the initial goal was to demonstrate the value of design through doing. With her team, they kicked off an eight-week engagement project and brought together a cross-functional team focussed on design. The team of twelve was comprised of people from marketing, product, IT, design and pricing. They were trained in basic design methods and ethnographic research and were brought to customer's homes to observe people in their day-to-day lives. It was an immersive and intensive process and in the words of Cecilia, they were 'absolutely blown away'. For many of them, this was the first time that they actually spoke directly to a customer. Not only did it give them an opportunity to connect with customers, they came away with a number of important insights on how their lives could be improved. Following a design process, they prototyped and tested a number of ideas, eventually ending up with three possible ideas to take forward. This team became the Connected Home team and they continued to work together for the next six months after the initial pilot project. They have really embraced a customer-centric mindset and changed the way they approach product development. This pilot project became the catalyst to kick-start design thinking in Telstra and provided demonstrable evidence of the value of design to senior management.

Cecilia's cause was further helped by the CEO and the Board of Directors' visit to IDEO. Hearing and seeing the results of a 'design thinking' approach highlighted what Cecilia was trying to achieve with her Connected Home team project. Leveraging the success of the project, Cecilia proposed the establishment of a design thinking centre of excellence to educate Telstra about design thinking and to work at redesigning the internal processes. On July 27, 2013, the Design Practice was established with the support of CEO David Thodey.

Stages of transformation

1. Strategic change programme started in 2009.
2. Repositioning design as a strategic and innovation tool through pilot projects.
3. Setting up the Design Practice in 2013 as a design thinking centre of excellence in Telstra to help embed design thinking in Telstra.
4. Helping teams be more customer-centric from ideation through to launch of a new product or service in order to deliver more creative solutions to customers.
5. Supporting Telstra's goal of becoming a world-leading technology company by using design as a framework.

What can we learn from Telstra's story?

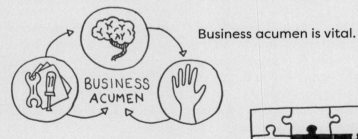

Business acumen is vital.

Aligning the value of design with organisational goals and strategies.

Measuring and evidencing value from the start.

The Design Practice

The Design Practice team offers three main services: 'Design-led Strategy' focuses on facilitated design thinking for strategy, 'Customer-led Design & Testing' helps teams design products and services, and 'Design Capability' service helps embed design in Telstra's culture and practices.

'There are three goals of the practice today: one is to help the business develop a design-led strategy. It's about adopting a customer-centric approach to innovation and using that to come up with new strategies. The second goal is to help teams use customer-centric design all the way through from ideation to launch of a new product or service. And then the third goal is to embed design into the mindset and the culture of Telstra. How do we transfer the designer skills back into the business? How do we empower the business?'
Cecilia Hill. General Manager, The Design Practice

The 'Design-led Strategy' service helps teams define strategy, propositions, products and services through customer-driven design. The type of questions explored includes: 'How do I use customer insights to innovate new processes, products and services?' It's essentially helping teams uncover customer insights and using those insights to spot opportunities and drive strategic priorities. The second service offered by the Design Practice is the 'Customer-led Design and Testing' service. It helps teams turn research insights into design ideas that can be prototyped and tested. They address questions like 'How do I design and test our products or services to deliver customer advocacy?' This is the sharper end of the product development stage and involves turning a research insight into ideas that are prototyped, tested and implemented. The third service offered is the 'Design Capability' service, which teaches customer innovation techniques and helps teams understand how to use design thinking. This involves running Design Thinking awareness events, Design Boot Camp training sessions and one-to-one coaching.

Setting up the Design Practice was a calculated risk. But it was a gamble that paid off. Since 2013, they have tripled their team to 30 people. They have their own dedicated premises and a strong internal as well as external identity. When they started out, they were focused on delivering five key projects with the product management group. Since then, they have worked with a number of different business units and have completed 56 projects between July 2013 and June 2015. They have a further 29 projects lined up for the first quarter of 2016.

Apart from working on projects, the team has also developed Telstra's design process, called the four Ds: 'Determine, Discover, Design & Test, Deliver'. They felt they needed to adapt existing design processes to one that fitted with the way the organisation is run. They have also developed a toolkit and have created a design thinking education programme that is now being rolled out. In the last few years, they have prototyped different ways of implementing design and doing design on different types of projects.

Framework maker

One of the key challenges in embedding design in a large and complex organisation like Telstra is to maintain momentum and continually gain the support from different teams and levels in the organisation. One of the ways they achieved this was to turn design from a mysterious 'black box' activity into a clear and understandable step-by-step process. It wasn't until design was articulated into the 4Ds: Determine, Discover, Design & Test, and Deliver, that they could make design understandable for people. The 4Ds created by the Design Practice team helped them communicate to the wider Telstra audience what design is and how design can help them with their problems.

Design's role in this instance acted as a *Framework Maker*–giving a clear process and structure for innovation to happen. The 4Ds performed a number of functions. They were a way to frame Telstra's approach to design. It was a framework that offered a structured approach to understanding problems and generating ideas to test. Design is also framed as a clear method to help the organisation and its people become more customer centric. The 4Ds also became a way to communicate how design is used in Telstra. Cecilia recognised the importance of making the 4Ds framework work with the other processes such as the agile and lean processes already in used in the organisation.

Power broker

Cecilia is acutely aware of the importance of aligning the role of design to the company's customer-centric principles. Design thinking is known for its humanising attributes and bringing a more customer-centric focus to an organisation. Design in Telstra is positioned as a key capability that brings together transformational activities already happening in the organisation and focuses it on customer needs. It helps align the organisation's work and concerns around key user needs, thus diffusing fractional internal priorities and help create a unify goal. Co-creation is also a useful way to break down silos and encourage people to work together.

'It creates this environment where people actually talk to each other. It creates an environment where people come together to generate ideas and look at problems and it really builds stronger relationships between people through collaboration.'
Cecilia Hill

Evidencing value and translating design for business

'The initial goal of the Design Practice was to demonstrate the value of design and to get the team up and running and settled. And we have achieved that. We have proven that design works.'
Cecilia Hill, General Manager, The Design Practice

One of the key contributing factors to the Design Practice's growing influence in Telstra is showing the value of design in a language that is understood by the business world. This is evident in a number of ways in Telstra's story. The first, and probably most obvious, way is to measure impact through metrics already used by the organisation.

At the start of each project, they will discuss and decide on the success factors. The team usually uses two types of success factors. The project-level success factor is generally quite well defined since it's based on a known customer problem. They use well-established metrics called the Net Promoter System[SM] that uses regular customer feedback to understand what customers value and what annoys them across all the activities that comprise the customer experience. The second type of success factor is based on measuring a change in the mindset of people before and after the projects. This is usually done for larger projects and levels of customer-centric awareness are measured before and at the end of the projects using qualitative questionnaires. The organisation has already been tracking change in mindset since the switch to a more customer-centric approach in 2009. Apart from using these known metrics, the Design Practice team also holds a review session after every new product and service launch to find out what worked, what didn't work and also how it is currently performing in the market. They also measure how likely it is that the internal stakeholder will recommend working with the Design Practice team to gauge how well they have done.

'The personal side of the transformation journey has been quite strong. I feel like we are now are an organisation that really cares about the customer. We have seen improvements in individual episodes and as well as at a strategic level but not at the same rate. So our challenge is still to ensure our transformation efforts are focused on changing the system, not just the interactions.'
Scott Barclay, HR Director, Organisational Change and Transformation Design.

A more subtle but much more significant way in ensuring design remains relevant to the organisation is using vocabulary that is understood by the business. It's important to align the activities with what's already in place and to really understand how it fits in the overall strategic directions of the company. This strategy manifests in a number of ways. For example the 4Ds framework is designed to 'speak' to people who have had no experience of design. The way the three main services of the Design Practice are named and described is designed to resonate with the type of activities and language already in used in the organisation. It's also really important that Cecilia's team develop a good understanding of the business. Having dedicated teams delivering in the three different service areas ensures that they not only develop the right competencies required for the difference services, they also start to understand the business through 'framing'.

'It is pretty crucial to be able to explain what design can do but to also have a business understanding of how the people are being measured and how the business operates. Having general business acumen is extremely important.'
Cecilia Hill

Building and sustaining a design-led community
The Design Practice has achieved a lot in its short two-year history. There has been a significant shift in how design is understood and used in Telstra. There are an increasing number of teams in Telstra asking for support from The Design Practice. A sign that design is becoming better understood is that many of these project requests are located in the realms of business design, developing new processes and programmes. For example the Chief Risk Officer has commissioned the team to look at designing a new framework for risk management. The Design Practice team has worked with the Human Resource team to develop a new HR intranet portal and mobile app to improve staff's three primary HR tasks of submitting leave, time sheets and accessing pay-slips. Significantly, the use of design and design principles are slowly being embedded into the way the organisation operates. Before a project gets signed off, there has to be a formal review to see if it is going to follow a design and customer-centric approach.

The number of projects that the Design Practice is involved in has also increased year-on-year. Transformational activities started by the organisation since 2009 have seen a much bigger focus on customer needs, and design has been really useful in helping the organisation achieve its aim. For example, there is evidence of prototyping occurring more often in projects. There is more collaboration happening between groups and a lot more co-creating activities. The word 'design' is being mentioned a lot more at different levels, significantly at board meetings. This indicates the increasing profile design is gaining in the organisation.

Now that design has gained a foothold in the organisation, the next challenge for Cecilia and her team is to ensure that there is a consistent approach to using design. Getting everyone excited about design is great, but ensuring it is applied appropriately in the right context is really important to ensuring its integrity. Design is not a panacea to all the organisation's ills. Cecilia knows it is important to start defining for the business when and where design thinking should and shouldn't be used. Everyone talks about it now, but running a workshop with post-it notes does not represent 'design thinking'.

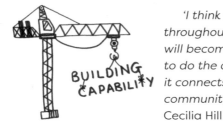

'I think now what we need is to find internal champions throughout the business that we can really train up and who will become designers within the business, not necessarily to do the design work, but to talk about design and how it connects to the business. And to create that design community throughout Telstra.'
Cecilia Hill

Building capability is only a start. You then need to build a supporting culture to enable people to act differently. For example, the general operating model of a company is to work on a cycle of a yearly budget. However new product development or strategic projects are rarely contained within a one-year cycle. So there is a need for a more flexible funding model.

For design to flourish and maintain its relevance to the business, it has to truly drive strategy. For example all new products and services should have a customer-centric focus and if they don't, then they should not be funded. Design shouldn't be confined within a centre of excellence, but instead be sitting at the same level of capital planning. Design competencies should be found in all levels across different silos and teams.

'Ultimately we want to have a mandate and a sign-off from the top to say design is one of the ways we are going to do things around here and this is how we're going to do it. So that's what I'll be working towards, to get that mandate, essentially.'
Cecilia Hill

What the Telstra example has shown us is that it is really important to frame the use of design with the goals of the organisation. Telstra's new vision is to become a world-leading technology company. Cecilia's and the Design Practice team's continuing challenge is to help Telstra realise this vision using design as a framework.

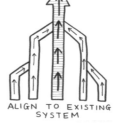

What are the conditions for impact in Telstra?

- Finding support and sponsors from the organisation at all levels.
- Understanding the business and its aims.
- Using an approach that can align to existing systems.

- Using language that the business understands.
- Evidencing and measuring impact through KPIs at the start of the project.
- Acknowledging and planning for a long-term engagement since transformation takes more than 5 years to achieve.

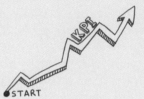

What have been the challenges so far?

- Design has historically been limited to usability and user experience in the organisation.
- Making the case for the value of design as a strategic tool and then continually demonstrating through outcomes.

What type of change still needs to be achieved?

- Ensuring a consistent approach to using design.
- Building design competences across different silos and teams.
- Establishing resources, structures and processes to support the new design-led culture.

US Department of Veterans Affairs: Giving veterans a voice

Challenges in innovation in the public sector

Compared to businesses, public services exist in an infinitely complex social system. Goals and values are not solely driven by a profit or growth model and as a result are much more ambiguous, difficult to quantify and generalised. Add in additional barriers such as: a risk averse culture, ad-hocism, short-term thinking, and lack of incentives to create a truly performance-driven and innovation-driven culture, all make innovation in  public services almost a herculean task to achieve. These are the challenges currently facing individuals and teams tasked with bringing in new innovative practices. While there are many parallels we can draw from our two public services examples in the book, the US Department of Veterans Affairs' story offers us a fascinating insight into how design is being used at the federal government level to bring about transformational change.

This story is especially pertinent as it illustrates the humanising potential of design to catalyse a change in a governmental culture that often prioritises process over people. Design's focus on the user enables it to diffuse tensions and realign teams around a common goal. By making the veterans' experience 'real' through visually compelling personas and customer journeys, VA staff were able to reconnect with what veterans want and need from VA services. The human-centred design process adapted by VA (Discover, Design, Deliver) acted as a *Framework Maker* to provide a road map for how to use design. While VA is still a long way from achieving a people-centred culture, design is being used to help catalyse this transformation through its humanising force.

///

Who we spoke to
Sarah Brooks, Chief Design Officer, Veterans Experience Office, VA
Secretary Robert A. McDonald, US Department of Veterans Affairs
Tom Allin, Chief Veterans Experience Officer

Why change?

The US Department of Veterans Affairs is the 2nd largest department in the U.S. Federal government behind the Department of Defence. Declining service quality, long waiting times to access VA healthcare, compounded by the increasing number of veterans accessing VA services as a result of the Iraq war, and ongoing conflicts have resulted in increasing pressure from Congress and the public to improve its services to veterans. A very public scandal in 2014, the result of a report by the inspector general regarding systemic failures at VA, became the turning point for the department to fundamentally question and change how they work

Design roles that enabled change in the VA

Types of changes achieved through design

Since 2014

Changing products & services

Changing organisation

Changing the process of change

What has a design-driven approach brought to VA?

- The personas and customer journeys have been a powerful focal point to help realign organisational values to the needs and perspectives of veterans and their families.
- Confidence that the solutions created respond directly to veterans' concerns.
- Measurable improvement in customer experience and trust.

Caring for Veterans

The US Department of Veterans Affairs (VA) is the second largest US federal agency, second only to the Department of Defence. It employs over 340,000 people and the annual budget in 2014 was $152.7 billion. The Department's mission as stated in their 2016 Functional Manual[1] 'is to serve America's Veterans and their families with dignity and compassion, and to be their principal advocate in ensuring that they receive medical care, benefits, social support, and lasting memorials promoting the health, welfare, and dignity of all Veterans in recognition of their service to this Nation'.

VA offers three categories of services to support veterans and their family in the US and US overseas territories: healthcare, benefits and memorial services. The Veterans Health Administration (VHA) is the biggest of the three areas and surprisingly provides the largest integrated healthcare system in the US, with 156 medical centres, 800 community-based outpatient clinics, 126 nursing homes and 35 living facilities. It doesn't just offer medical care but also cultivates ongoing medical research and innovation to improve the lives of veterans. The second administration area is the Veterans Benefit Administration (VBA) and their role is to oversee benefits and offer some social services that include the GI bill, employment help, disability compensation, vocational rehabilitation and home loans. The National Cemetery Administration (NCA) runs 147 national cemeteries and provides memorial services. 19,000 acres of land are devoted to memorialising those who have honourably served the US.

In total, services and benefits are provided through a nationwide network of 144 medical centres, 1203 community-based outpatient clinics, 300 Vet Centres, 56 Regional Offices, and 131 National and 90 State or Tribal Cemeteries[2].

Crisis breeds Innovation

'Crisis breeds innovation', these were the words of Secretary Robert McDonald when we interviewed him. A crisis offers a foundation for change, and change was needed at VA. It isn't easy running any large organisation, be it private or public. In the case of the VA, it was made more challenging for a number of reasons. Firstly, like many public services, the VA is a 'highly brittle service' that is hampered by top-down bureaucracy divided into silos. Services have been designed with processes rather than customers in mind and, like many other public organisations, VA is risk-averse and rule-bound. Secondly, ensuring adequate service is delivered to over 22 million veterans (about six per cent of the entire population of the US, almost a third of the UK's population and almost the entire population of Australia) is itself a huge challenge. This is further compounded by the ageing population of Korean and Vietnam War

Stages of transformation

1. Robert McDonald was appointed the Secretary of Veterans Affairs in June 2014 and immediately aimed to improve the service quality and delivery by creating a new programme called MyVA.

2. Human-Centred Design (HCD) was being championed within the VA Center for Innovation (VACI) by Amber Schleuning, an Army veteran. Amber recruited Mollie Ruskin, a Presidential Innovation Fellow to conduct design research on the current state of customer perceptions with VA services in the winter of 2014. In September 2014 Sarah Brooks, a Presidential Innovation Fellow was recruited to continue the human-centred design work.

3. Secretary McDonald appointed the first Chief Veterans Experience Officer–Tom Allin in Jan 2015. The Customer Experience Office was set up and led by Tom Allin, Sarah Brooks, Julia Kim (also a 2014-2015 Presidential Innovation Fellow) and Walt Cooper.

What can we learn from VA's story?

Innovating in government has additional challenges that need to be clearly identified and addressed in a way that is specific to that context.

The humanising potential of design offers a powerful incentive and unifying force in a government environment that often prioritises accountability rather than delivering better services.

Design historically is perceived as a 'neutral' and apolitical profession. In a government environment that can be politically charged, this perceived 'neutrality', through its focus on the citizen as the ultimate reference point, can be used as a powerful tool to enable change to occur.

veterans as well as the large influx of wounded Iraq and Afghanistan veterans in the past 14 years.

Working in a government agency undoubtedly brings other challenges. While there are a lot of dedicated people working at VA, they have been hamstrung by numerous constraints. For example having to work with out-dated IT systems, hierarchical processes, legacy projects as well as having to work around the government's funding cycle. All these factors combined to create the perfect storm that exposed how ill-equipped VA was to deal with the changing needs of veterans. VA reached a crisis point in 2014 when the then secretary Eric Shinseki resigned in the wake of a report by the inspector general regarding systemic failures that led to delayed care and falsified records (unknown to Shinseki) at the agency's hospitals. VA is still suffering from the fall-out of this scandal and in a sense, even though the crisis was very public, it was cathartic for everyone involved at VA and became a rallying point for internal change within the agency.

Embracing customer experiences

The change at VA has come in a number of forms. Firstly it started with the appointment of Robert McDonald as the new Secretary of Veterans Affairs. Secretary Bob (he likes everyone to call him Bob but in such a formal environment, people prefer to address him as 'Secretary Bob') is a West Point graduate and had a distinguished career for 30 years at Proctor & Gamble (P&G), one of the world's largest multinational consumer goods companies. Secretary Bob was appointed P&G's Chief Operating Office in 2007 before becoming its CEO in 2009 and eventually retiring in 2013. He came out of retirement to take up the role of VA Secretary because he was keen to serve veterans. His appointment was a signal of intent from President Obama. Not only was he not an ex-military man, he was someone who has spent all his life working in the private sector. This clearly indicated to the public that the President wanted a change in VA's approach, one that is about adopting best practices, introducing innovations and ultimately one that delivers better results.

Coming from P&G, which has had a long history of human-centred design and design-driven innovation, Secretary Bob knew that the key to changing the organisational culture was to focus on people rather than process. This meant focusing on customers' needs to drive all future development and change at VA.

The first thing Secretary Bob did, when he took over as Secretary, was to begin a comprehensive effort to transform the agency into a customer-focused one by creating the MyVA initiative. MyVA is part of a larger strategic plan to reorganise the department to ensure success, guided by ideas and initiatives from Veterans, employees, and stakeholders. The aim of MyVA is to provide Veterans with a seamless, integrated, and responsive customer service experience. And while the MyVA initiative provided the VA with a single regional framework to enhance services, it has also been used to communicate the change in VA's approach to delivering personalised services for veterans.

Bringing in new practices

A few months after Secretary Bob started at VA, Sarah Brooks began her one-year residency at the VA Center for Innovation as one of the 2014/2015 Presidential Innovation Fellows. The Presidential Innovation Fellowship is an experimental programme set up by President Obama 12 months before his first term ended in early 2012. The programme was initially conceived as an entrepreneur-in-residence idea and was aimed at bringing innovation into the public sector by pairing  top innovators from the private sector with civil servants and change makers in federal government to tackle some of the nation's biggest challenges. It is a 12-month programme where the Fellows are embedded within a federal agency to collaborate on challenges with innovators inside government. The White House recruited a range of talented innovators from the private sector who would use lean, agile and design methods in projects of national importance. In total 18 fellows were appointed as the first-ever Presidential Innovation Fellows. Due to its initial success, further rounds of fellowships were announced and appointed. Sarah was appointed a fellow in September 2014 and was placed in the VA Centre for Innovation to work with the team driving radical transformation.

Using the human-centred approach as a framework to understand design

The VA Center for Innovation is a team of innovators and doers within VA who are dedicated to driving innovation. In 2009, then Secretary Eric Shinseki established the VA Innovation Initiative in order to help VA transition into a 21st-century organisation by tapping into new sources of ideas from the existing staff as well as bringing in innovators from the private sector. The initiative later became formalised as the VA Center for Innovation (VACI) in 2010 and has since worked to identify, test, and evaluate new approaches to the VA's most pressing challenges.

Amber Schleuning, an army engineer veteran and VACI's Deputy Director was particularly influential in bringing in a human-centred aspect. She brought the first Presidential Fellows into the VACI in 2013 and has championed the HCD approach ever since. In 2014, VACI ran a project to pilot the use of human-centred design to understand America's veterans. The project was lead by VA's 2013-14 Presidential Innovation Fellow, Mollie Ruskin and she planted the seed of HCD in VA. The HCD process piloted and now used by VA comprises of three stages: 'Discover', 'Design' and 'Deliver'. The 'Discover 'phase consists of research, synthesise and define, while the 'Design' stage comprises of ideation, prototyping and testing. The final 'Deliver' stage involves refinement, building and implementation of the idea.

Sarah Brooks started working at the VACI in September 2014 and her first project was to conduct ethnographic research to uncover customer understanding and their journey. This led to the publication of the 'Voices for Veterans' in November 2014, which identified seven key personas representing the VA's customers. This project was really influential in demonstrating the value of design to different people at VA and epitomises the value of the HCD approach. To this day, the project and its outcome are often used as an example to champion the value of design and the HCD approach within VA. The HCD process offered Sarah and her team a simple process and language to introduce and explain design to other VA staff. It also gave the organisation the confidence that it could achieve its goals and ambitions, despite the fact that the precise nature of the outcome is not known at the beginning of the process. The visual nature of the personas offered tangible markers of progress and insights, which helped to create a pragmatic, purposeful conversation, which in turn helped catalyse a change in culture from process to people.

Catalysing and accelerating change

From the moment Secretary Bob arrived at VA, he could see that the entire operations was process rather than customer focused, function rather than purpose focused. He wanted to flip the mindset from an inward looking perspective to an outward facing perspective and to focus on understanding a customer-first perspective, which very much fits into the humanising capability of design. Not only did he have the foresight to create the MyVA initiative to drive this transformation, he also knew he needed a new position at the highest leadership level that would act an as the accelerator for that change.

At the start of 2015, Secretary Bob hired Tom Allin (a former CEO in the food industry) to be VA's first Chief Experience Officer and to lead the Customer Experience Office. The role was to change the way VA approaches its customers in order to deliver better services. It was a significant move as it was the first Customer Experience Office established in a federal agency that had a cabinet level appointee. Although VA was not the first federal agency to establish a Customer Experience Office, it had the largest customer base, as the other team located in the General Service Administration (GSA) are only one-tenth the size of VA.

What are the conditions for impact in VA?

- Innovation in government has to be driven and supported on a number of fronts and at different levels. The White House Presidential Innovation Fellowship programme, The US Digital Service and 18F are evidence that design can enact positive change in the federal government. Code For America is the strongest example of design enacting positive change at the state level.

- In order to use design to enable change in an organisation, you need a head of the organisation (for example Secretary McDonald) who not only champions its cause but also understands what it means to be truly human-centred.

- Strategic placement of the design team that sits across functions and departments so as to be seen as a neutral resource.

- Establishing a design leadership position that would act an as the accelerator for that change.

- Resilience and patience to innovate in government.

What have been the challenges so far?

- Connecting backstage changes (how employees feel, think and behave) to front stage experiences of customers.

- Overcoming organisational habits that don't serve employees or veterans and their families.

- Overcoming government processes and structures especially in the areas of HR, IT and Contracting in order to bring designers and innovators into government.

- Securing discreet funding for the Customer Experience Office in order to be less vulnerable to a change in administration.

Secretary Bob knew that in order to transform customer experiences, he needed an office and a team that would traverse the three main administrations dealing with healthcare, benefits and memorial services. As a result, the Customer Experience Office is the only organisation within VA that is looking completely laterally across these three areas. The team reports directly to the Secretary and is charged with customer understanding and ultimately the creation of seamless end-to-end customer journeys. This positioning is important as the Office is seen as a neutral entity and this allows it to broker relationships across the different areas, which is important if they want to completely transform the end-to-end customer experiences. Design is being deployed as a neutral *Power Broker*, one that seeks to leverage the independence of the profession of design and its focus on the user as the ultimate reference point.

The Customer Experience Office's core team consists of Tom as the leader, with Sarah Brooks leading the Insight and Design team, Julia Kim as the Chief of Staff, and additional directorates around Measurement and Performance Improvement, Omni-Channel experience and Field Operations. The office has been growing rapidly and now has around 20 people based in Washington and with the aim of growing the field offices to 180 people around the US. The team in Washington works closely with partners across each of the three administrations and the VA Center for Innovation (VACI), using Human-Centred Design (HCD) process to drive customer and employee insights. They are focused on ongoing ethnographic research about their customers' wants and needs, and using some fundamental tools of design such as customer journey maps and personas to communicate the insights from their ethnographic studies. Apart from establishing the Washington team, the aim is to set up local Customer Experience teams in the district offices in New York, Atlanta, Chicago, Denver and Las Angeles in order to have more direct influence at the site of service delivery.

Humanising the veterans' experiences

One of the key roles design has played in this story is the role of *Humaniser*. Above all else, the focus on people and humanising the veteran's experiences is the key driver in catalysing a culture change. This is achieved in a number of ways. We already spoke about the impact the 'Voices for Veterans' publication has had in the organisation. In addition to these very visible insights, the Customer Experience Office uses a co-creation approach when working with internal staff.

Sarah Brooks leads the Insight and Design team within the Customer Experience Office. Her team is currently 15 people strong and their role is to help the organisation uncover customer understanding and how to respond to those needs. Her team is pushing HCD across the organisation in a number of

ways and primarily engage by offering themselves as internal consultants. There are generally three ways in which Sarah and her team engages with the wider VA community.

Sarah calls the first type of engagement as 'co-created generative conversations' where her introductory session on HDC becomes the starting point in exploring what their needs are and how these approaches might help them. These conversations might then lead to developing projects with the team to help them address problems they have identified and taking them through a HCD process.

There are a few high-priority agency initiatives owned by the Customer Experience Office, such as the new platform for service delivery, vets.gov (owned in partnership with VA digital service), and the consolidation of VA's call centres. Sometimes Sarah's team are asked to look at emergent problems; for example, Secretary Bob requested the Insight and Design team to look at the disability claims process after receiving a number of emails from Veterans. The team conducted discovery research to understand what was happening and to identify key insights that frame the next design phase. A workgroup was put together and one of the designers from Sarah's team was embedded within it. The designer conducted a series of co-design workshops with that group to figure out what interventions they wanted to make and what subsequent work streams were going to come out of that work. She was dipping in and out during the project, offering guidance on the HCD process and keeping connected with them as they carried out their work across the different work streams. It was a lightweight consultation and was designed that way. It's important that the ownership of the project stays with the workgroup and that they make decisions supported by the Insight and Design team. This ensures a higher level of involvement and support for the eventual ideas that get implemented.

'Great customer experience requires great employee experience.'

This quote by Sarah hits the nail on the head. Anyone involved in improving customer experiences will know that it can only be delivered by employees who not only understand customers but who also have the tools to help them deliver a good customer experience.

VA is an unusual agency in that 30 per cent of its employees are veterans themselves. So, in theory, they should be in tune with what a veteran's experience is and also be highly motivated to want to help veterans. And yet staff are often surprised by the VA's customers' perception of poor service. Why is this so? A majority of the employees have been working at VA for 10-30 years and, as a result, many of the processes that seem obvious to them are often confusing to customers. They also start to lose the outsider's perspective of the service and this often leads to a disconnect between what they think they are delivering and how it's actually being received by the customer.

One of the ways the Customer Experience Office is trying to address this disconnect is by creating a 'relationship map' to help link the activities happening 'Back Stage' at the VA with the 'Front Stage' of actual customer experience. Why is this important? Firstly, employee experience exists back stage and is currently invisible to veterans. Similarly customer experience exists on the front stage and is currently invisible to VA. 'On Stage' is where employees and customers interact. Getting staff to understand the links between back stage and front stage will allow them to develop a different view and understanding of the relationship they have with veterans.

Part of ensuring a great employee experience is to provide the VA's staff with the appropriate training, resources, incentives, tools and support to enable them to deliver better services. These include the development of foundational tools such as veterans' personas, developed by Sarah Brooks when she was the Presidential Innovation Fellow at the VA Center for Innovation, to understand the different kinds of customers they serve. Additionally, detailed journeys of veterans' life-stages were carefully researched and mapped out, along with the ten key customer journeys through VA. Both the personas and the journey maps have been extremely well received. Staff are especially drawn to the emotional power that personas can offer, not only do they communicate the key needs and attitudes of different types of veterans' experiences, they also offer rich insights into the different contexts of life after the military. Both sets of tools have helped kick-start a lot of conversations at different levels and generated interest in design as an approach.

Challenges in catalysing a culture change

VA is a huge organisation and enabling culture change to permeate from the top to the lowest level of the organisation is a huge challenge. While the leadership principles used in business are the same ones used in government, the context and how it is executed differs significantly. Secretary Bob talked a lot about the difficulty in creating the inseparability of interest between the organisation and the individual. In government there are fewer tools and ways to leverage this. The fact that most government employees stay in the same jobs for an extended period of time tends to mute their excitement or their ability to contribute at a high performance level. It's just hard for anyone who has been in the same job for twenty years and doing things the same way, to change. For this reason, Secretary Bob has established a three-day training programme called 'Leaders Developing Leaders' where the expectation is that the individual employee comes out of the process with a project they have proposed to complete in a hundred days that will result in a better outcome for veterans. It is also a way for each successive layer of leaders to train the next subordinate level and cascade that training. So far 15,000 leaders have already gone through the training.

Programmes like Leaders Developing Leaders help prepare current VA staff for change. However, in order to grow the design capability internally, the Customer Experience Office has had to bring in new people to drive the change process at the different field offices. The office is setting up local Customer Experience teams in each of the five VA districts and are looking to hire 180 people (110 in the field offices and 70 in Washington) to deliver improved customer experience. People often talk about the 'triangle of hell' when having to navigate government structures: IT, Contracts and HR. Not only has it been extremely challenging to recruit new people (it has taken 11 months to hire 15 new people), it has also been challenging finding the right people. There have been historically no roles for design in the US government and this meant having to write new job descriptions and titles that speak to and attract designers.

'Government is not set up for effectiveness or efficiency. It is set up for fairness and to prevent fraud. So with all their checks and balances in place it's almost impossible to get things done.'
Tom Allin, Chief Experience Officer

Acknowledging that it would be a herculean task to change the structures of government, finding ways to circumvent the system seem to be the most effective and quickest way to innovate in government. The Innovation Presidential Fellowship was designed to introduce this disruptive element. 18F was set up by the White House and the GSA to spread best practices adopted from the most successful technology companies. They are essentially the US government's in-house digital service consultancy and service any governmental agency requiring support.

Bumpy road to change
'The acid test for me of any culture change is when the lowest level person on the totem pole or on the hierarchy chart knows what they do every single day and how that connects back to the vision of the organisation. The analogy I like to use is when President Kennedy was touring one of our space centres, and he asked the guy pushing the broom what he was doing, and he said I'm putting a man on the moon. When the person pushing the broom knows they're putting a man on the moon you know you've succeeded.'
Secretary Robert McDonald

Everyone we spoke to acknowledged how difficult it is for an organisation the size of VA to change. However, Secretary Bob, Tom and Sarah were incredibly positive and buoyed by the fact that they have witnessed noticeable change, assuring them that they are on the right track. For Sarah, she has observed a growing understanding of customer experience and increasing interest in HCD as people are exposed to its theories and methods. More people are using vocabulary from HCD and are also comfortable explaining it to other people. And while this is only the start, it does offer early evidence that design can catalyse culture change through a clear focus on peoples' needs and deep empathy as a means of approaching sensitive, cultural challenges.

VA and the Customer Experience Office have only just started on their transformational journey and still have a lot to do. The most immediate pressure is to 'get as much done as possible' before the next election-cycle where a change in administration might undo all the good work they have done thus far. It's important for the Office to achieve proof of concept and evidence that design does have a significant and measurable impact directly on improving veterans' experiences. For that reason having the right kind of measurable impact such as veterans' satisfaction and trust is important to track progress made. Ultimately measurable progress only matters when the veterans' experiences change for the better. The team at VA are quietly optimistic that they are moving towards achieving this goal.

Notes

1. www.va.gov/ofcadmin/docs/VA_Functional_Organisation_Manual_Version_3-1.pdf
2. Statistics as of August 2015

What type of change still needs to be achieved?

- Strengthening the impact of the Customer Experience teams in the five field offices across the US.

- Showing measurable impact quickly and before the Presidential transition.

- Establishing an organisation-wide understanding and application of design to help them become more customer-centric.

SAP: Humanising technology

The start of an IT giant

SAP started life as a German software company set up by five former IBM employees in 1972. In its first year of operation, it employed nine employees and generated roughly $350,000 in revenue. Since then, it has grown to be a world leader in enterprise applications software and is the third largest independent software manufacturer, behind Microsoft and Oracle. It currently employs around 75,600 employees with office locations in more than 130 countries. Their annual revenue is a staggering $19.3 billion and they serve over 296,000 customers in 190 countries worldwide.

The Design Thinking journey in SAP

Hasso Plattner, one of the five original co-founders of SAP has been instrumental in bringing design thinking into SAP. Design thinking resonated with Hasso because it offers a more human-centred approach to problems. He felt that their connection with their customers had been lost as they expanded over the years. He also felt that the time was right for the world of business and software development to put a more human emphasis on their work.

As a result, he sponsored Stanford University's first design school, now more popularly known as the 'd.school' aimed at bringing design thinking to the business world. Concerned that good ideas were often being lost through a lack of an innovative culture, Hasso set up the Hasso Plattner Institute (HPI) at Potsdam, Germany in 2008 to provide 'a better' education for software engineers. The School of Design Thinking was later established at the HPI Potsdam to introduce design thinking to the curriculum. At the same time, Hasso started to introduce and accelerate design thinking into SAP by bringing in 35 design thinkers to create the Design Services Team, a multi-disciplinary

Who we spoke to
Jochen Guertler, Senior Design Strategist, Design and Co-Innovation Center, SAP

Why change?

SAP is the world's leader in enterprise software. Although they have been extremely successful and dominant in this sector for a number of years, they felt that their connection with customers had been lost as they expanded over the years. They were very good at what they did but had become too engineering and process focused. They wanted to reconnect with customers and felt it was time to make people their focus again.

Design roles that enabled change in SAP

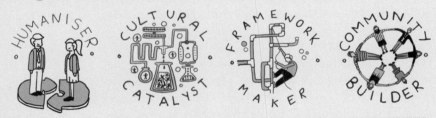

Types of changes achieved through design

Since 2008 - Internally in SAP
Since 2012 - Externally through the Design and Co-Innovation Center

Changing products & services

Changing organisation

Changing the process of change

What has a design-driven approach brought to SAP?

- Given teams a sense of 'agency' due to direct exposure to user concerns.
- Developed empathy not only for customers but also with team members.
- Fostered a more open, collaborative and user-focused culture.

group that is housed in the CEO's Office. This enabled the group to reach across the organisation to use design to impact SAP strategy, products and people. The team initially comprised 35 individuals from diverse backgrounds that worked with different teams in the organisations to introduce design thinking.

Setting up a visible and tangible team ensured that design became a strategic priority to drive innovation across the organisation. This model of creating a 'design thinking' focal point either through the creation of a team, space or an innovation lab has been copied and replicated in many of our book's examples, for example in Telstra.

Since the creation of the Design Services team, design thinking has slowly spread, effecting not only internal processes and culture but also becoming evident in the organisational structure. Sam Yen initially joined SAP as a senior member of the Design Services Team. He then led the SAP AppHaus, an innovation team tasked with building new solutions, establishing new markets, and reaching new users for SAP. In June 2014 he was appointed SAP's first ever Chief Design Officer to further cement the strategic role design plays in driving innovation in SAP.

Using design as a Framework Maker

The initial Design Services team was set up centrally and tasked with spreading the use of design thinking in SAP. However, after two years it was evident that for design to be truly embedded into SAP, there had to be access and engagement with design at a local level. Hence various incarnations of the Design Services team were set up in different departments to provide a range of services internally and externally. Since 2013, a new service team was set up as the Design and Co-Innovation Center (DCC) to elevate perception of SAP's user experience and design. DCC became the focal point for SAP to deliver UX design services to SAP customers.

DCC offers a portfolio of services that helps in the areas of user analysis, design thinking, design, proof-of-concepts, custom development and training. Their services are loosely grouped into four areas: Advise, Innovate, Empower and Realise. The 'Advise' service helps business to identify, evaluate and explore ways to improve an organisation's UX strategy. The 'Innovate' service helps organisations to develop a competitive advantage by creating value for their customers. The 'Empower' service supports organisations in achieving the best user experience. And finally, the 'Realise' service helps customers put their UX strategy into action by supporting them in optimising, customising and adapting existing SAP software specific to their purpose. Examples of DCC's work includes: designing user interfaces, working with customers on co-innovation projects, coaching customers in design thinking, developing custom

Stages of transformation

1. Hasso Plattner, one of the original co-founders of SAP, founded the Stanford University d.school with David Kelley (2005).

2. A branch of the d.school was set up in Potsdam, Germany. 35 design thinkers were brought in to collaborate with the corporate strategy group to make design thinking a strategic priority at SAP. The aim was to drive innovation across the organisation (2008).

3. Various design thinking teams were set up across the organisation to work at a more local level with different departments (2010).

4. SAP made a strategic decision to change the perception of business software through human-centred design (2012).

5. Design and Co-Innovation Centers established in various locations to offer design thinking approaches to external clients (2013).

6. SAP's first Chief Design Officer appointed (2014).

What can we learn from SAP's story?

Organisational change only happens if individuals experience it for themselves.

Learning through doing is the most impactful way to enact change.

Turning a new practice into a 'new normal' requires assimilation and eventual ownership of the practice.

applications in existing SAP software, developing mobile apps and customising graphical user interface in SAP software.

The team at DCC comprises professionals from 27 countries, working as design strategists, UX designers, user researchers and programme managers. The team works out of four locations: Heidelberg, Berlin, Palo Alto and Seoul. The DCC team's working style is both agile and scrum-based and highly iterative in its engagement with end users.

Since its inception, DCC has had over 500 customer engagements and has reached thousands of end users during a variety of user research activities. They use design thinking as a *Framework Maker* to develop empathy for customers and users by creating tangible reference points such as personas and user journeys. They collaborate closely with customers and in some cases establish design practices within customers' organisations through hands-on learning and prototyping to ensure they have the ability and capacity to deliver the strategy.

DCC uses a design thinking approach with three key stages: Discover, Design and Deliver. In the 'Discover' stage, the DCC team would scope out the work, help their client explore the problem space and then conduct appropriate user and market research. As in all stages the DCC does this in a very interactive workshop-driven way involving customer and potential user groups from the beginning. They then synthesise their findings, turning them into ideas and prototyping them with their customers. DCC usually starts with simple paper prototypes co-created with their clients, which are then used to validate the ideas. To help them refine and communicate the idea further, the paper prototypes are turned into wireframes and interactive prototypes. Implementation during the 'Deliver' stage is sometimes handled by the DCC team but most often is handed over to their client's internal IT team or other development teams within SAP.

Making data more meaningful

How do you leverage existing statistical data and use it to create a better experience for your customers, the ice-hockey fans? This was the question the German Ice Hockey League (DEL) brought to DCC with the aim of enhancing the fans' experiences. DEL manages all the first division games, organises marketing activities and promotes young talent. DEL had already collected a lot of statistical data about recent games but

had not really utilised the data apart from using it internally (for example to announce a player's birthday).

To help answer this question, DEL needed to identify and understand who their fans were and what they were interested in. Two innovation coaches from DCC ran a two-day co-innovation workshop with members of DEL and their media partner to introduce them to various design thinking tools. Part of the activity in the workshop involves speaking with a range of stakeholders and fans to help the team develop a range of fan personas. This activity helped the team realise that a majority of the fans are not ice hockey experts and mostly attended games to enjoy the family-like atmosphere. This insight led the team to develop a dashboard interface idea that would display interesting facts about each team. It would enable fans to compare different kinds of statistics for fun (for example number of fouls or assists) but also help them understand the sport better. Paper prototypes were created during the final stage of the workshop and used to test with users.

A second workshop was held to explore how the content can be expanded and defined. At this stage, other experts from the DCC were brought in to help realise the vision. Visual designers were tasked with developing the interface design and thinking through the navigation workflow. DCC was also involved in the final implementation and realisation of the dashboard and leveraging existing SAP software to help DEL improve its data analysis capabilities.

The public dashboard went live in the new season and hockey fans are now able to access data in a visual and fun way through the dashboard. They can now track statistics in real time, for example passing accuracy and player scoring ratio. The dashboard has been really successful with fans and has increased fan engagement with DEL across its entire communication platform – Facebook, website and Twitter.

Design as a change management tool

Increasingly the DCC is working on projects using design as a catalyst for organisational change. Organisation X (anonymised) is a world-leading food and beverage company creating well-known products in the global market. DCC has recently started to work with Organisation X to help them become more innovative, agile and future proof. To create the right momentum for the design topic, the DCC conducted several 2-day workshops with a total of 500 employees of Organisation X. Design is being used as a *Framework Maker* – introducing design thinking processes and methods to the organisation and offering them tools to help them innovate.

The second aspect of the engagement involved DCC running a 'Train the Trainer' programme, where they spent 20 days training future design thinking coaches in the organisation. DCC initially trained five coaches

LEADER TRAINING

and these coaches have now subsequently trained other members of the organisation. This enabled Organisation X to grow competencies internally and build shared practices.

Design in this instance was used as *Community Builder*, through its hands-on and action orientated workshops. It helped build shared practices. This in turn has enabled the creation of different innovation communities in the organisation. As a result, different teams in the organisation have initiated a number of new projects. They range from redesigning aspects of the intranet system to developing a new app to help maintain customer loyalty around a product range. DCC were only minimally involved in these projects but they have learnt that the most persuasive way to demonstrate the effectiveness of design thinking to an organisation is to apply it to projects that have defined focus and produce concrete, measurable outcomes (for example reducing the number of steps required to fulfil certain tasks). DCC also felt that it was perhaps too early for the internal teams to apply design thinking without prior experience. After the initial buzz and excitement, producing concrete outcomes is really important to ensure the continued use of design as well as offering continued project support after the initial training stage.

Personal story of transformation

It is evident that SAP and Hasso Plattner have really pushed for the adoption of design thinking through the establishment of the initial centralised Design Services Team and its various local incarnations like the Design and Co-Innovation Center. While these were very explicitly and overtly public actions to embed and promote design thinking, it's important that we bring our focus back to the individual and explore how the transformation feels for an SAP employee. We spoke to Jochen Guertler, who works as a senior design strategist at DCC. He joined SAP in 1998 and has first-hand experienced of the transformation as well as seeing it through the eyes of his colleagues. Over the years, he has worked on a number of software projects as a developer, software architect, product owner and team leader. He joined DCC's Heidelberg office in 2013 and his current role is to organise and moderate design thinking and co-innovation workshops and projects with clients and to guide them through their innovation journeys. He also supports customers to help them develop more innovative and user-centred solutions.

He was a developer and a software architect when design thinking was first introduced to SAP. Since design thinking was introduced in 2008, he has seen 'extreme changes' in SAP. While he acknowledges that SAP is a massive organisation and that the rate of change differs and is inconsistent across the different teams, he has seen and experienced a significant enough shift

What are the conditions for impact in SAP?

- A strategic aim driven by leadership.
- Being pragmatic in how people learn about design to ensure a positive experience and subsequent adoption.
- Continuous support and access to required resources.

What have been the challenges so far?

Using design thinking with clients

- Ensuring deeper understanding and application of design thinking through doing.
- Getting access to end users, a key part of a design thinking approach.
- Using design as a change management tool requires expertise from change managers.

Embedding design thinking in SAP

- Uneven application and adoption across the different teams due to the size of the organisation.
- It's a long-term process requiring the right support and organisational structure.

What type of change still needs to be achieved?

- Further cross-organisational adoption and alignment (across global teams and offices).

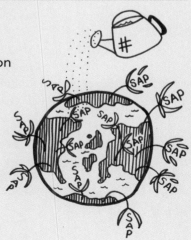

in how people work to convince him of design's effect as a *Cultural Catalyst*. For example teams are working in a more iterative manner, working in shorter developmental cycles, weeks rather than years. He has also observed that people collaborate better and are more comfortable with teamwork. They are also much more user focused and devoting much more of their time to talking and working with clients collaboratively.

'Realistically it's not always easy to gain access to users, but since the overall perception is that users are important, we have to find ways to talk to them, however difficult.'

Teams are also more comfortable showing unfinished projects in the form of first and rough prototypes at an early stage to clients to get their feedback and to learn whether the idea could work for them. Teams have adopted the mantra of 'fail early to succeed sooner'. Compared to how a traditional software engineering process works, this is indeed a huge shift in the mindset and behaviours of the staff. The physical spaces in the SAP offices have also changed. There are now a lot more creative and flexible spaces available for teams to sit and work together, both internally and also with customers.

'Traditionally in big companies like SAP, there are normally never ending discussions about everything. While it's important in some cases, in many cases it's more important to act. Our attitude is becoming less about talking and more about doing.'

Jochen also observes how attitudes towards design have changed in SAP, especially amongst the software developers. Previously, design was merely thought of as adding to the visual aspects of the user interface. However, they now see design as a holistic engineering solution and as playing an important part in creating the overall experience of their product. He sees this attitude as part of SAP's DNA now.

For Jochen, the change did not happen overnight. Of course he was aware of the drive by Hasso and the executive team to push design thinking. However, it only really clicked for him when he was given the opportunity to be involved in a concrete design thinking project. In 2010 Jochen moved from a development role to a research role and he started work with the German

Sailing team on an innovation project. They wanted to use design thinking in the project and colleagues from the Hasso Plattner Institute were asked to train and support them in using design thinking. It was a revelation to Jochen–for example, he learnt so much by talking to sailors and their coaches during the Kiel week, a big sailing event in the north of Germany. Prior to that, although Jochen was told it was important to speak to users, he was never given the opportunity to do this first-hand. After this experience, he realised how powerful yet simple this approach was. It was also the first time he experienced a multi-disciplinary team working with design, marketing and business experts. This project and experience completely changed the way Jochen viewed and understood what design can do.

Jochen believes that if you have this positive experience with this new way of working, then there is a very good chance (provided it is continually supported) that you will apply it to your daily job. He also believes that design thinking is not only useful for innovation projects but suited to incremental development, which is what a majority of the SAP teams are involved in on a daily basis. Small changes like faster testing cycles with key users will make a huge difference to the way they work.

'This is not a mind-blowing fact. True change only happens if you try it and experience it for yourself. This was a key aspect in how design was introduced and embedded in SAP. Hasso really pushed for it and SAP invested a lot of money to provide training and supported projects using design thinking. Although not all projects were successful or implemented in the end, it did not matter since the key benefit was that people had the opportunity to experience what design can do.'

Humanising aspect of design
One of the most important roles of design in SAP is to act as a *Humaniser*– challenging the techno-centric way of working and enabling a closer connection to consumer needs. For example, Jochen ran a project retrospective session after the end of a project. During the session one of the developers in his team remarked that this was the very first time in his 10-year career in SAP that he truly understood the reasons why he had to implement certain features in the software. And it was the first time he had direct contact with users right from the start. He was really engaged in the project and was highly motivated simply because he understood the needs of the users and why it was important to them. He was able to see the direct impact of his work on them. This was extremely powerful and often taken for granted in terms of someone's agency in a project.

The collaborative nature of a design thinking approach not only requires working closely with the client, it also requires working closely with the project team. Continuously sharing and early prototyping of ideas builds a shared understanding of the issue and the possible solutions. It is increasingly about fostering not only empathy with end users but with your own project team members.

'I see higher motivation amongst colleagues. I see many examples of people working together and there is a better working environment. Ultimately all organisations are made up of people in the end, therefore the people have to change if you want to change the organisation.'

Prototyping is considered one of the key tenets of design thinking and often talked about as a very powerful engagement tool. For example, creating quick paper prototypes has helped the DCC project teams to communicate ideas with clients at an early stage. This not only establishes a shared understanding of the project, it helps motivate and enthuse clients right from the start. It is no longer seen just as an IT project, but one that is tangible and real for the team to see its effects on actual users.

Sustaining a cultural change

It's almost a decade since SAP started its design thinking journey. Design has not only transformed the way people work in SAP, it seems to have transformed its DNA. They have provided us with a rare example of how a culture change of this magnitude can be achieved and, importantly, maintained.

One of the challenges of maintaining a cultural change (and this can be seen in the Telstra example) is ensuring that there is consistency in the support for design in the strategic direction of the company. This has had to come from the executive team to ensure that new members buy-in to the original vision that Hasso had from the beginning.

Another key condition in maintaining this open, collaborative and user-centred culture is to ensure that the team constantly reflect on how they want to work. Now that a design-led culture is established and has become 'normalised' it is important to keep challenging and questioning whether this is the most appropriate way to work. This will build ownership of the process and links back to Jochen's earlier point of enabling people to develop agency in their own work. It will also lead to the most transformative aspect of design, influencing their change process.

And finally, it's really important for the organisation to develop its own 'version' of what it means to be design-led and to decide how it fits within their existing corporate culture. It has been really interesting to explore how design has had an impact on the process, structure and culture at SAP. However, the establishment of the DCC and the work they are doing shows how much SAP has made design thinking their 'own'.

Adur and Worthing Councils: Using technology as driver of change

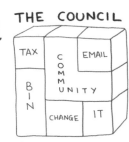

Introduction

The Adur and Worthing Councils case study is a prime example of using design as a *Technology Enabler*. Digital transformation was required to overhaul their out-dated IT system and rebuild both the back end and front end to support a more integrated and efficient service delivery. Why was this needed? When 10 per cent of your budget is dedicated to maintaining your IT system, it is easy to understand why the council was focused on overhauling its existing system as a way to reduce cost and improve efficiency. And, in order for the digital transformation to be successful, there needed to be a significant design component attached to it to ensure it met user needs.

The wider ambition of the council is to use this digital transformation as a way to challenge our understanding of the role of the government, citizens and service providers. The digital transformation agenda was the anchor point to help them start this process. They are doing this by implementing a new digital strategy that is based on the idea of 'Government as Platform'[1] using a design-led approach. It involves building and delivering user-focused digital services that can eventually be applied across different councils. It is about moving away from large monolithic IT systems designed for a single departmental use to one that is cloud-capabilities based. It has become the backbone to radically reshaping the way they work and simplifying services they deliver.

Who we spoke to
Paul Brewer, Director for Digital & Resources

Why change?

Adur and Worthing Councils were looking to transform their IT systems into a digital service that not only reduced costs and improved efficiency but also transformed the relationship between the council as the service provider and the users. Design, in effect, was there to make sure the right kind of technology was being used in the right way. At the same time, Adur and Worthing Councils have a more strategic and longer-term ambition to embed human-centred design into how they think about, design and deliver services.

Design roles that enabled change in AWC

Types of changes achieved through design

Since 2014

Changing products & services

Changing organisation

What has a design-driven approach brought to AWC?

- Human-centred approach.
- Experiencing change through doing.
- Prototyping quickly and producing exemplars.

'We're talking about really questioning how we operate, what's the role of the citizen, what's the role of the community, what's the role of volunteering, what's the role of community and voluntary sector in achieving this outcome.'

Government as platform

'The platform model is about going away from the vertical line of business systems and onto a horizontal platform which extends beyond the organisation.'

There are number of compelling reasons for councils to rethink their role and how they operate due to the public sector spending squeeze in the UK since the 2008 financial crisis. It has become widely accepted that the role of local government will increasingly change to one of a facilitator, commissioner and curator, rather than just a service provider.

Adur and Worthing Councils are situated in the south coast of England and actually made up of two separate councils[2]: Adur and Worthing with a combined population of around 160,000 people according to a 2011 census. The council is responsible for providing services like rubbish collection, recycling, Council Tax collection, housing and planning applications for the local area. Rapid growth in certain sectors and changes in local demographics are putting pressure on the council to change the way they do things.

Adopting a 'council as a platform' model requires an approach to technology different from what has been the norm in government. Historically IT systems designed for local government were bespoke, delivered by large IT suppliers on long-term contracts. They are standalone software and do not connect nor overlap with other software used for different functions. However, the reality is that while parking and council taxes are two separate services, they both share similar requirements. For example, they both need a payment system–the same person might be unfortunate enough to pick up a parking fine and at the same have to pay their council tax. It's not surprising that there are currently 440 databases that the council needs to access on a daily basis.

It is not hard to see why an overhaul of the current system is required. But how can this be done?

Stages of transformation

1. Fundamental review of technology led to the creation of the Digital and Change Strategy.
2. Setting up the Design and Digital Team gave a focal point of change.
3. Digitising systems and processes led to more collaborative and open ways of working.
4. Redesigning user-centred services with the aim of making them easier to use, having a common interface and saving cost.

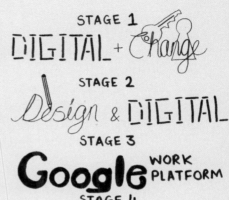

STAGE 1
DIGITAL + Change

STAGE 2
Design & DIGITAL

STAGE 3
Google WORK PLATFORM

STAGE 4
New Services

What can we learn from AWC's story?

GET ORGANISED

Ensure that the organisation is ready for change.

Clear out the 'gubbins'– spend time sorting out the mess before beginning the transformational journey.

Getting the right people to champion and embody the change is crucial.

FINDING THE RIGHT PEOPLE FOR THE JOB

Design and Digital Team

In 2013, the new Chief Executive of Adur and Worthing Councils set out his vision and plan through the Catching the Wave strategy. The strategy outlines the aim to redesign services so that they are citizen-centred, easy to use and digitally enabled. They believe that by being genuinely user centred and working with a network of partners and contributors outside the organisation, new, more personalised and accessible services will be delivered in new ways. They started by hiring Paul Brewer as the Director of Digital and Resources in 2014. He was tasked with implementing this strategy over a three to five year period.

Paul has had many years experience working in local government, and previously spent six years at Brighton and Hove Council as the Head of Performance in Children's Services. As well as setting out a more clearly defined implementation plan, his first six months were focused on preliminary work to ensure some basic conditions were in place for the next stage of transformation. They undertook a fundamental technology review with the help of a London-based consultancy, Methods Digital and developed a new, funded strategy to move away from a line of business systems and adopt flexible cloud-based platform technologies that they can design and build. Paul knows from experience that without getting the underlying technology foundation and architecture right (the 'plumbing' as he calls it), all other digitally driven innovation projects will fail or have minimal affect in the long term.

To realise this new digital strategy, a Design and Digital team was created to provide service design, insight and open data services, digital services and project management support. The team has a number of functions. Firstly it's to support the 'digitisation' of the organisation through redesigning internal IT services and processes. Its second function is to help redesign services delivered by the council. Its longer-term aim is to eventually provide support to external organisations looking to redesign their services. It is a design-led team, adopting a people-focused design approach to support the organisational, community and economic growth challenges identified by the council. A core team was put together by Paul based on a skills audit and additional team members have been drawn from different departments. In addition to this, two new members of staff were hired, the Head of Design and Digital and a digital service designer.

'The Design and Digital team works like an agency within the organisation and will be absolutely obsessed with training the managers of services in the required performance management and project management skills. As far as possible we will have project leaders that are from the service in question and we'll support them to make the change so they own the change.'

Technology enabler

The first stage in this transformation is what Paul terms as 'digitising the old world'. For staff to understand what 'digital' means, it was important that they experienced working in a new way enabled by cloud-based technology. Staff emails and calendars were moved to the Google for Work platform. Google for Work is a service from Google that provides customisable enterprise versions for several Google products, which are based on web applications; examples include Gmail, Docs, Drive, Groups and Hangouts. The advantage it offers over desktop-based software is that documents, calendars and emails can be accessed on any computer, and collaborating and commenting on shared documents is a lot easier and more transparent.

Design is used as a *Technology Enabler* ensuring the organisation selected the right set of technologies to support a new way of working. In Adur and Worthing Councils' case, it was trying to help them be more collaborative. Paul was particularly keen for invited stakeholders to contribute directly into a shared document since it opens up the possibility of added value coming from unexpected places. Its also reduces the chances of documents being duplicated when being worked on by different people. This move has also allowed the council to save money on  productivity software licenses and more importantly to help staff rethink what a 'document' is, how they work and perhaps, what their work really is.

The team continues to work on a number of internal projects as part of the digitisation plan and one example is to redesign their performance management system. The current system requires manually collating and compiling project updates into a report for senior management who may or may not have time to read them. Instead of focusing on getting a report completed, they have decided to move away from a document centric environment into one using a free online project management system called Trello. It's a live project environment set up to track four areas of performance measures and updates of each project can easily be accessed and linked to more detailed Google Documents. There is a simple traffic lights system that shows progress and can easily highlight any problem areas.

Through this simple and cost-effective redesign, it has demonstrated to the organisation how a cheap digital technology can have a significant cultural impact. Many teams in the council are already using Trello to manage their work in a more visible and collaborative way.

Capabilities and user driven

Paul is adamant that they are change agents and not technologists. That said, without getting the core technology right at the start, it would be very difficult to enact any meaningful changes. Their approach has been to focus on creating core capabilities such as booking, reporting, paying, and case management that can be used across different services. This not only saves money but is also a flexible system with a common interface across different services and has the added benefit of reducing support costs.

Traditionally, government services have always been delivered through a transactional model–described as a vending machine model by Tim O'Reilly. People pay money in and expect a service to be delivered. In contrast, a platform model is akin to a bazaar–a place where different vendors (large or small) participate in an ecosystem that is flexible and open. Vendors can come and go in a bazaar depending on demand and hence knowing what a user needs is important. As a result user journeys from different stakeholders have become important starting points for service redesign.

A long-term aim of the council is to create a platform that is not only used for the delivery of its own services but is also used by other councils. If other organisations choose to deliver services through the platform, then this shift represents a move beyond a technology project and into changing the way councils support local, civic, community and business activities.

Organisational readiness

Organisational readiness came up repeatedly in our conversation with Paul as he drives and manages this ambitious transformation. Paul understood the challenges and conditions required for change to happen, gleaned from years of experience working in less than ideal conditions. He was adamant that this scale of organisational change must not be left to chance and that anyone looking to embark on a transformational programme has to ensure the right conditions are in place.

Having the support of the executive team is important but having the ability to uncover and trace hidden power structures is far more useful. Paul was also fortunate to be in a position where he had access to financial, human and legal resources through his portfolio. This condition is key since without the power to control how financial and human assets are redistributed and channelled, it would have been difficult to invest in the right people and technology required for the plan.

What are the conditions for impact in AWC?

- A clear vision and rationale for change, supported by commitment from Senior Management to achieve the change. They also need to invest money and time in the process and expect visible change after 12-18 months, and not sooner.
- Assign a dedicated team to drive and manage change, and populate it with people with the required skills and mindset.
- Team must have access to financial, human and legal resources in the organisation
- Rebuilding the foundations – clearing out the 'gubbins' – existing problems before transformation can begin.
- Adopting an appropriate transformation strategy– in Adur and Worthing's case, it was an evolutionary rather than revolutionary plan to start by digitising the way they work.
- Using early project exemplars to prove value of design quickly and immediately.

What have been the challenges so far?

- Changing a risk-averse culture to one that embraces risks.
- Moving from a vertical line of business systems and onto a horizontal platform that extends beyond the organisation.
- Building a design-led community from scratch.
- Overcoming hierarchical structures and political agendas to recruit team members across the organisation.

What type of change still needs to be achieved?

- Changing the way services are delivered and managed.
- Changing culture (attitudes & behaviours) towards public service delivery.

Culture change would certainly fail without team members who are motivated, skilled and have a 'can-do' attitude to champion and drive change. One of the most difficult challenges for Paul has been to evaluate who has the skills and drive required for this experiment. Paul took his time observing and communicating with different people in the organisation to gauge their 'readiness' levels. At the same time, he had to be decisive in removing people who have been disruptive and obstructive to the plan.

'The biggest challenge that's been really hard–is to judge who's in and who's out. So who's going to have to leave the organisation... that's the constant strain because what I want to do, is to say hey, we're one big family, we can all do this together.'

Getting the dynamics of the Design and Digital team right is only the beginning. Moving the entire culture of the organisation requires a leap of faith. The move to the Google for Work platform introduced staff to new ways of working that aren't too radically different but that were still capable of challenging existing assumptions of 'how things are'. It has helped allay anxiety and mistrust of internet-based technologies specifically around information security.

'An important condition to enable this digitally driven transformation was around the preparedness of the organisation to accept more risk. We were quite lucky because the information security person at Adur and Worthing was very open to seeing things differently. And if he weren't, it would have been a priority for me to address.'

And finally, the 'proof is in the pudding'. This popular English saying reflects the reality of many of the transformation stories that we have heard. It is very important to show evidence of success early on in a transformation project, and probably more so when using a lesser-known design-led approach. Being able to personally experience positive change is extremely valuable and must not be underestimated. The Design and Digital team is doing this by using different projects to establish and try out different capabilities and by implementing these changes gradually.

'We are creating exemplars which allows us to establish these capabilities by doing real projects and doing things one step at a time.'

Challenges

Managing change is challenging–without a doubt. A recurring theme from stories in this book is about finding the right people who are 'up for it'. Selecting the right people internally and externally are the fundamental building blocks to a transformation project. Having a free hand to overcome existing hierarchical structures is key to enabling this. It is also critical that the appropriate foundations (structures, people, process, finance for example) are put in place before work begins. 'It is about sorting out the 'gubbins', a term often used by Paul when talking about this issue. One of the early challenges for Paul was being the lone voice calling for change in an organisation unused to a design-led approach. This often improves over time when a change team has been created to champion and deliver change, but it's fundamentally about being strong-willed and the confidence to trust in the process.

Impact

24 months into the transformation project, the council have successfully established their new platform and have won two local government digital awards in recent months. They have 'fixed the plumbing' meaning that their digital services are being built using reusable capabilities and common data. This has led to shorter development delivery times, for example they recently designed and built a service to support the homeless (housing options) in two months, from discovery to implementation, with a significant change in process for the service user.

Engaging effectively with users

It's a well-accepted fact that there is a growing divide between government organisations and the people they serve. Encounters with governments are often described as 'patriarchal' and even 'patronising'. There seems to be a genuine fear about knowing how to engage properly and, in doing so, of opening up a can of worms. There is also a tendency not to take feedback seriously or not to consider it representative of the wider public opinion.

For Paul, design is important as a *Framework Maker* because it gives him the tools to not only engage effectively with users, but also gives him the confidence that it will lead to better services. He needs people who are confortable doing user research and using those insights to drive and direct possible solutions. The additional challenge is to ensure that skills are seeded across the organisation so that everyone will be user-focused in whatever role they are in. And finally, having a design process is not just about meeting user-stated requirements; it's also used as a *Power Broker* to reconcile multiple voices and specialisms for the benefit of creating a more human-centred local government for the 21st century. It's ultimately about challenging everyone (citizens, community, staff and businesses) to think about public services differently.

Where next

Paul and the executive team at the council know that they still have much to do when it comes to embedding human-centred design and using design as a *Framework Maker*. Although their technical developer capability is now strong, they are still running 'digitisation' projects in the main rather than developing new services. The executive team is conscious of this and have plans to address and assign more budget resources. Paul is finalising a case for a 'civic lab' which will help them develop a toolkit and skills for human-centred design, and they have a set of exciting projects they will run through it, including rethinking housing repairs (a more radical strand alongside the urgent digitisation they are doing), the future of (government) work, rethinking car pooling (with HiyaCar), and crowd-funding civic space (with SpaceHive).

Notes

1. Government as Platform is phrase coined by Tim O'Reilly in a 2010 chapter featured in the Open Government book. It has been adopted by the UK's Government Digital Services and describes a common infrastructure of shared digital systems, technology and processes.
2. Local government in the UK are referred to as local councils. There are different types of local councils and each has a responsibility for a particular range of local services.

Satellite Applications Catapult: Creating an innovation culture from the ground up

Design and space

Design as an innovation and change tool has, until very recently, never played a huge role in the space sector for a number of reasons. Firstly, it's a highly technical industry dominated by an engineering process. Secondly, the projects have previously been on a vast national scale commissioned by governments and the public sector. But like many industries, the ongoing commercialisation of the space sector has meant that new customers are coming from the private sector looking to exploit space technology for the mass market. This is where design comes into its own and offers a more customer-led approach that captures the imagination of the market and offers the space sector an approach to tap into these new opportunities.

Design in this example is utilised as a *Technology Enabler* for its ability to translate highly-specialised technology through visualisation and prototyping activities to make it understandable for a non-specialist audience. At the same time, it acts as a *Framework Maker* for the space sector looking to find ways to identify opportunities and turn them into new products and services. The turn towards users and customers brought on by the change in market focus has also meant that the design as a *Humaniser* role has given Satellite Applications Catapult (SatApps) and their clients ways in which to

Who we spoke to
Dan Watson, Head of Design and User Research
Joel Freedman, Senior Design Thinker
Stuart Martin, CEO
Antonia Jenkinson, Chief Operating Officer/Chief Financial Officer

Why change?

The space industry in the UK has massive potential to grow and new market opportunities are opening up, especially in the satellite sector, due to an increase in commercialisation and affordability of technology. Satellite Applications Catapult (SatApps) was set up to take advantage of space infrastructure to deliver new services and applications and maximise the economic potential of the UK's space industry. Design is seen as a way to bring much needed innovation into the sector. SatApps offers an example of how challenging it is to use design as an innovation and change tool in an organisation and sector that is still relatively new to design.

Design roles that enabled change in SatApps

TECHNOLOGY ENABLER · FRAMEWORK MAKER · HUMANISER · CULTURAL CATALYST · FRIENDLY CHALLENGER

Types of changes achieved through design

Since 2013

Changing products & services

Changing organisation

What has a design-driven approach brought to SatApps?

- It provided the tools to help them become more user-centred.
- They achieved a higher success-rate for new products and services.
- It has helped shorten delivery times and enabled SatApps to have a quicker route to market.
- It helped diversify their organisation into innovation consulting.

understand customer needs. Additionally, design has been a *Cultural Catalyst* to SatApps internally and to some extent, externally with project partners and networks by bringing new approaches to working specifically around the generation of ideas and working more collaboratively. A design approach has also transformed the modes of communications by using visualisation to help people understand and communicate SatApps' offering both to an internal audience as well as to others in the space sector.

Changes in the satellite sector

One of the fundamental changes that is happening in the space sector is the entry of commercial companies into what has been historically a state-sponsored space industry. Space exploration requires a huge amount of resources and was only feasibly funded through state agencies like NASA, the European Space Agency (ESA) or Russia's space agency, Roscosmos. However, commercial companies like SpaceX, Orbital ATK, and Planet Labs are beginning to challenge the dominance of state sponsored space technologies. SpaceX, owned by American entrepreneur Elon Musk, designs, manufactures and launches advanced rockets and spacecraft. The cost of launching satellites has been falling due to an increasing number of companies like SpaceX offering commercial payload delivery to orbit. SpaceX is also pioneering a reusable launch system development programme that can be reused many times, thus lowering costs further. Due to the commercial nature of these companies, the need to have a more define market and to show value and a return on investment in this sector has become much more acute. Additionally the new requirements for quicker intersatellite links (via the Space Wide Web) will allow satellites to open up a whole range of more timely applications, which they were previously too slow to address.

Historically it has been difficult to respond to market demands in the space sector quickly and in an agile manner due to the high level of complexity, risk and costs involved, all resulting in long development time. However with launch and satellite manufacturing costs falling coupled with hardware becoming smaller, smarter and cheaper, new opportunities to innovate have significantly increased in the last few years. For example maritime surveillance, which requires a constellation to deliver reasonable revisit times, is now possible because the space industry's relevant products and services are becoming much more affordable.

The sector is highly specialised and technical but is increasingly becoming more multi-disciplinary. However, solutions and concepts are still presented as highly technical and lengthy reports, which make it difficult for non-specialist audiences to understand. As a result, fundamental questions that should have been asked at the start of the project are sometimes missed due to the highly technical knowledge required to understand the technical specifications. While

Stages of transformation

1. The establishment of SatApps in 2013 with the intention of using design as a strategic function.
2. Changes in management and internal growth meant that design, as a strategic function, was not fully established till the middle of 2014.
3. The design team gaining prominence and establishing its worth to SatApps.

What can we learn from SatApps' story?

It is challenging to use design as an innovation and change tool in an organisation and sector that is new to design.

It's often difficult to bring design-led innovation to a field that is highly technical, specialist and specification driven. User-centred perspectives are often ignored unless written by someone with a technical background involved in the project.

Multi-disciplinary teams bring with them differences in perspectives but also differences in working processes that are difficult to overcome without a concerted effort.

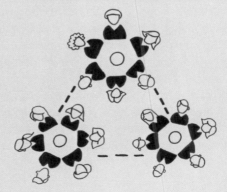

non-specialist team members and clients are unable to question assumptions for the same reason. This also has a knock-on effect when it comes to trying to present and sell ideas to non-specialist companies looking to exploit the use of the new satellite applications.

When the reliability of a product is more important than its usability, it has meant that user needs are generally a secondary concern. For example, SatApps was approached to help a company improve the design of their software developed specifically for a satellite scheduler project. The software enabled the operator to track and identify the satellite passing overhead. However it required the operator to read through a very long manual before s/he was able to use the software. This is often a typical case within the satellite sector, where limited user research is done to inform the design of the product.

Satellite Applications Catapult

Satellites have become a critical part of our lives. At any given time, there are as many as one thousand active commercial, military and scientific satellites in orbit helping us collect and distribute vast quantities of data and giving us a detailed picture of what is really happening in our world. Satellites are used for a range of purposes, from helping us navigate using a GPS enabled smart phone, monitoring wind velocity and  pollution levels to enabling communications between different parts of the world. Globally the satellite market is expected to be worth £400 billion by 2030 and the UK government has committed to capturing 10 per cent of that market share (£40 billion). Its potential for growth is enormous and, with long standing expertise in satellite manufacture and data applications, combined with an entrepreneurial outlook, the UK is well placed to take advantage of the changing landscape ahead.

Based in Harwell, near Didcot, the Satellite Applications Catapult (SatApps) was established in May 2013 by Innovate UK (UK's innovation agency) as one of a network of centres to accelerate the take-up of emerging technologies and drive economic growth through the exploitation of space. It is part of the Catapult network, set up and partly funded by the UK government to focus on nine areas of possible economic growth in the UK economy. Catapults are technology and innovation centres where UK businesses, scientists and engineers work side by side on research and development.

SatApps helps organisations make use of and benefit from satellite technologies by bringing together multi-disciplinary teams. Although SatApps do offer some discreet products and services, they are designed to act as a catalyst to help companies exploit new technologies and support development of new products to markets. The organisation has expanded rapidly since it was established and now has over a hundred employees.

Most of SatApps's activities are grouped in distinct programmes of work. This structure enables them to focus on areas where significant economic impact is possible while making the most of their funding. Some of the programmes include Intelligent Transport Systems, Blue Economy (marine environment), Sustainable Living, Explore and Government Services. SatApps also offer services designed to support companies in a number of ways and across the different themes. Services include: Knowledge Exchange, Research and Development, Business Support, Business and Design Sprints, Commercialising Ideas, Accessing Satellite Data and Market Reports.

The Design Team at SatApps

The Design team is part of the Business Innovation team and was set up to offer a more human-centred approach to SatApps and its clients. They currently have five permanent members with different expertise in product, interaction, architecture, graphics, UI/UX and service design. The team works on a range of projects spanning hardware, software, experience, service and communication. Their remit is to bring innovative, disruptive thinking to the organisation, encourage a culture where this is acceptable, and support projects through rapid-prototyping and human-centred design processes. And while SatApps is not the only company in the space sector to prioritise design and to use design to help it innovate, it is somewhat unique in using design as a strategic function to help them define direction and bring a more user-focused perspective to their industry.

Design was always meant to play a crucial role in the Catapult Network from the very beginning. One idea discussed was to create a separate Design Catapult centre when the network was set up. However, it was felt that design would have more impact if design capabilities were embedded at the different centres and for them to run and learn in parallel from each other. In reality, it has taken a lot longer for design to be established at the different centres. Although Satellite Applications Catapult was one of the first centres to establish a design team as part of the Business Innovation domain, it took the team 18 months of hard work to establish themselves and to identify key areas of service and value. As a result, design is now being considered a separate domain at the same level as the other seven skill domains.

The design team has been instrumental in raising the profile and advocating design in SatApps. Dan Watson is the Head of Design and User Research and leads the design team at SatApps. He has a background in product design engineering and was a member of the original design team when SatApps was formed in 2013. The other key member is Joel Freedman who works as a Senior Design Thinker at SatApps. He has a background in product design and mechanical engineering and his role at SatApps is to apply design as a process to problem solving across a full spectrum of projects across SatApps's programmes of activities. This includes using design for various purposes, from improving communication through better information design, the design of better products, services and systems through to using design strategically to design businesses and influence policy.

Design as a Cultural Catalyst

Design has not only emerged as a key function for SatApps, it is slowly but surely having an effect on its culture. Having a dedicated space for collaboration has marked out the design team as (positively) different, and embodied the different process and approach they bring to projects. The 'Design Cave' is a term used by all SatApps staff and keeps the design approach at front-of-mind (even if it's not entirely understood by everyone yet). It's also in a very visible location, being directly opposite the main office rather than being hidden away. Like so many of our examples in the book, establishing a physical space to enable creative and collaborative activities to take place should not be underestimated as a tool to encourage interaction. The space also provides a safe space to generate ideas that is helped in part by the use of the Design Cave when working at SatApps's offices in Harwell.

The design team collaborates with a number of internal SatApps teams, offering support in different ways. For example they support the Business Innovation and the Technology teams to engage and work with various industry partners and customers in the development of new ideas. Some of the work also includes supporting the teams in identifying user needs and user experience as well as prototyping ideas to test. They have also worked with the Marketing team to develop a more effective communication strategy through the 'Satellites for Everyone' campaign. The campaign was created to help demystify the industry and make it more accessible to external audiences. They have also worked to support and improve internal communications between the different teams to ensure that people with different expertise understand and agree on key project aims. The design team has also worked with the Applications Development team to embed user requirements into design features and specifications.

DESIGNATED
R
E
A

What are the conditions for impact in SatApps?

- Understanding the business context and sector requirements and how design can bring value is vital to generating credibility and respect from people in the industry.

- Having project advocates at the executive level enables access and continued implementation and evolution of using design as a strategic function.

- Having a dedicated space for collaboration and project work creates a different atmosphere and becomes the focal point of a creative culture.

- Having the right people in the design team that are not only able to work with people from different disciplines but also understand the sector and business.

What have been the challenges so far?

- The space sector is a highly technical and specialist field, which can mean a resistance to change and new approaches.

- The technical nature often results in slow development time making it difficult to adopt an agile approach.

- It's highly technology and engineering centric rather than user centric.

What type of change still needs to be achieved?

- Scaling activities and growing the design team.
- Fully embedding design as a core strategic function in SatApps.
- Increasing engagement and influence with the satellite sector in the UK through more external projects.

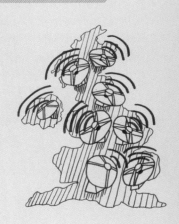

'The way that we communicate about ourselves is completely different from the way that normal space companies communicate and present themselves. And this is absolutely due to the design team. Others have contributed to it but it's the design team who really captured who we are and has made our voice much more natural and meaningful to sectors outside of the space community that we want to be engaging with.'
Stuart Martin, CEO

Design as a Technology Enabler and Framework Maker
The design team has begun to work increasingly with external companies and members of the Catapult network through activities such as the Business Design and Design Sprints. The concept of sprints is similar to Google Venture's Design Sprint methodology, which usually involves a five-day process used to answer critical business questions through design, prototyping, and testing ideas with customers.

The design team runs Business Design Sprints with Small to Medium Enterprises (SMEs) to identify opportunities and helps them transform new ideas into market ready products and services. The five-day process (distributed over a few weeks) involves combining business expertise with design thinking methodology. The team uses a design process as a *Technology Enabler* for their clients in order to help them formulate, explore and understand the value of new ideas, and how they can be exploited. In comparison to the Adur and Worthing Councils example, where technology was used as a platform for change, SatApps uses design to help clients understand and exploit technology.

They turn to that business understanding to identify new opportunities and distil ideas into a coherent pitch, backed by strong assumptions and a model that is understandable to investors and/or customers alike. They help teams generate innovative business models around their ideas and support them to a point where they are ready to communicate that idea to customers or to pitch for investments from potential partners.

A lot of what happens in Business Design Sprints involves a large element of translation–mainly describing the value of the business to someone outside of that field. This would often involve trying to simplify and visualise the message for a non-specialist audience and tailoring the process to suit the needs of the company.

While the Business Design Sprint service is focused on helping businesses develop and communicate a new concept (or pivot an existing product/service), the Design Sprint is aimed at helping companies create new user experiences and refine existing ones. The Design Sprint encourages companies to focus on solving specific design problems within a user journey, so that by the end of the process they are able to produce testable prototypes.

In both of these types of activities, design is deployed as a *Framework Maker* to enable people to work through an explicit process and create a common language to discuss ideas. It also offers the organisation the confidence (and a psychological safety net) that it can achieve its goals and ambitions, despite the fact that the precise nature of the outcome is not known at the beginning of the process. This is particularly important for companies who are new to design and the use of design in this way.

Challenges of introducing design

Examples from the book have shown how difficult it is to enable cultural change and SatApps is no different. It took a long time and lots of persistent work from the design team and a strong advocate at executive level to help SatApps understand the value of design and where it could fit. When the Catapult network was first created, a Design Special Interest Group study was conducted by Sam Adlen (Head of Business Innovation at SatApps) to help SatApps define its role as a catalyst and how design can be used to support it.

'Are we designing satellites or are we designing services? How do we fulfil the role as a catalyst? How do we help people develop new businesses and services out of new ideas? How do we make sure we are advising them to focus on the right direction so they make the most of the opportunity and ensure scalability and sustainability?'
Joel Freedman, Senior Design Thinker

While the intention to use design strategically to support innovation was present from the beginning, it took time for design to be used in that way. SatApps is made up of quite a lot of different teams, and each group has their own way of working. So while teams were encouraged to engage with the design team, many teams had limited understanding of design and were only using design where they thought relevant, which is mostly at the end of a development process.

Challenges establishing design as a Cultural Catalyst

Using design as a strategic and catalytic tool has been a challenge for the design team to develop due to a number of reasons. Initially the rapid growth of the company from essentially five to a hundred people over eight months and the introduction of new management slowed the process. This meant that projects developed during that period were often delayed until the management structure stabilised.

A second challenge came from the fact that many of the teams come from multi-disciplinary backgrounds. While this is usually beneficial to an organisation, in SatApps's case, it meant that they also brought with them existing assumptions in terms of processes and mindsets. As a result the design team found it very challenging to get the teams to work together initially.

While there have been challenges within the organisation and sector to accept the use of design as a catalytic force, the design team has also faced internal challenges linked to new skills required to operate in this new space. Both Joel Freedman and Dan Watson mentioned how important it was for designers to understand the business context of the projects. It is important for designers to have the skills and knowledge required to develop a business case for a new idea and to identify where value is added. Just as design thinking is crucial to business thinking, business thinking is equally vital for designers to create sustainable and desirable products and services.

'The marriage between the Business Modelling team and the Design team is really important especially from an innovation point of view. Although the terminology from design and business are different, they are very much referring to similar things. For example, designers see problems to be solved. Business people see opportunities.'
Dan Watson, Head of Design and User Research

'I realised that there are so many parallels between the different processes in business and design. For example the business analyst would focus on trying to find out how much people would pay for a product or service or they would want to work out distribution channels. In order to do that, you need to understand your users, you need to know about their lives, and this is just like the user profiling that designers do.'
Joel Freedman.

Making design work

It was really important for the design team to develop a good understanding of the sector and to adapt their processes to make it understandable and accessible. The adaptation, or the 'reformation' as Joel and Dan described it, is really important and can be really powerful when applied to business principles.

It was important for the design team to demonstrate their value by working with as many teams as possible and helping them experience what design is. Designers are extremely good at turning initial research insights into believable prototypes. This ability often became a powerful tool for the design team to illustrate a concept early on in a project in order to achieve buy-in.

The design team has also been very fortunate to have a prominent advocate in Sam Adlen who is on SatApps' executive management team and now heads up the Business Innovation team. He, alongside other champions such as Stuart Martin, the Chief Executive Officer, have always been key proponents of design thinking at SatApps. Having support at the board level has certainly helped the design team make the case for design thinking across SatApps's vertical and horizontal structures.

Offering people access to a different space where they can come together to work collaboratively with other people is an often-used trope in stories of innovation. SatApps is no different. The establishment of the 'Design Cave' with wall space to make ideas visible for reference and critique is an important step in establishing a design culture.

That said, all these conditions would not have helped if other teams did not like working with Dan and his team. While stating the obvious, being personable and having the ability to interact with people has been really important in helping to enable a design culture to be accepted and welcomed into SatApps. In this case, the humanising capability of design refers as much to people leading the change as to using design to humanise processes.

Changes in practices

The continuous work put into changing the organisation's perception about design has begun to show dividends. The design team has had to work very hard at communicating the value of design as an innovation and strategic tool. It has taken them close to 18 months to establish their team and activities and to make the case for using design at the start of every project. The design team is now much more involved in projects from the beginning and requests for support are focused on helping teams develop ideas rather than simply to communicate and package final products.

Another example of how the perception of design has changed in SatApps can be seen in a change in the project process. Before the start of a project, teams in SatApps have to review and respond to a list of questions. Questions around cost, collaborators, expertise and project length have to be discussed and determined before projects are given the go-ahead. In addition, management now requires each team to have consulted the design team before the start of any project. This ensures that the design team is used as a 'critical friend' and acts as a *Friendly Challenger* to help challenge assumptions and to draw out the idea's value right from the start. What helps is the transparency fostered by the use of design tools and approaches like service blueprints, deep and rich ethnographic narratives presented in user personas, role-play to demonstrate potential interactions and emphatic customer journeys. All these tools support a clear framework for challenging the status quo.

What is also encouraging for the design team is to note how often project teams return and request for additional support during the project. The design team is receiving a lot more request for support and participation compared to 12 months ago. This provides the strongest evidence to Joel and Dan that the use of design is gaining critical mass and teams are beginning to see the value of design in their work.

Scaling and engagement

Going forward, the design team is looking at ways to scale their work. There are a number of challenges for SatApps from a strategic point of view–one is focused on how to scale as a business and still have impact. The second is how to create a coherent narrative around impact based on the number and range of projects happening at the same time. Having over ninety projects running at various times in the organisation requires a strategic overview of what these projects are and what they are trying to achieve. Ultimately the projects have to contribute to the overall vision of SatApps rather than be simply seen as revenue streams.

Design is being used to help SatApps explore and develop a strategic viewpoint. This would involve activating a number of design roles, for example as a *Friendly Challenger*, as Dan and Joel work with the executive team to identify what is currently working and what isn't working so well. Additionally they will need to use design's role as a *Humaniser* in order to develop deep insights into the market and to understand users' perspectives to help them determine their key goals. They are focused on identifying where the real opportunities are and how to grow them.

In terms of expanding and scaling their activities, in the last few months the SatApps team has started to facilitate engagements with larger industry partners as well as smaller ones. For example they have conducted a 'Big Ideas Day' with a leading aerospace manufacturer. They have also recently been applying their expertise through the broader catapult network, deploying the same methodology used in their business design and design sprints.

The design team is also thinking about how they themselves need to evolve to meet the changing demands of the organisation. They are talking with other organisations like the UK's Government Digital Services to understand how they are embedding design. Currently SatApps has a central design team that goes out and works with the different teams rather than being always embedded in a particular area. In comparison, other organisations have designers that are embedded into the different teams. There are pros and cons to each model but, for the time being, the SatApps design team prefers to work across a broader spectrum of teams and to keep their identity as a team distinct.

Just doing design

The Satellite Applications Catapult story has shown us that establishing design competencies in a relatively new organisation can often be as challenging as introducing design in an already well-established company. Issues of rapid growth in a changing sector have meant that the design team at SatApps had to patiently establish their role in the organisation. And, although the SatApps's design team has made significant strides in the last 18 months, both Dan and Joel acknowledge that there are more challenges ahead. The team is already working on more strategically-focused projects and they want to continually expand in that area to ensure they deliver wider and deeper impact both internally in SatApps as well as externally with the Catapult network and other organisations. They also have the ambition to organise and disseminate their learning into a set of design and innovation tools created specifically for the space and satellite industry. At a more prosaic level, both Dan and Joel expressed the importance of getting to the point of 'not needing to explain design to everyone' and simply just 'doing design'.

Innovation Studio Fukuoka: Building community through citizen-led innovation

Introduction

Innovation Studio Fukuoka (ISF) is a citizen-led innovation platform set up and supported by the Fukuoka City Directive Council in 2013 with the help of Re:public, a Japanese 'Think and Do' innovation company. The aim of the platform is to use citizen-led innovation to drive the development of new business ideas in the city of Fukuoka. It is essentially a platform to create businesses rooted in the community.

This case study is somewhat different from the other case studies in the book, as it does not feature an organisation per se. However, ISF is essentially made up of individuals brought together for a specific goal and has used design in a number of ways to enact change in society. For this reason, this is an interesting case study for us to explore. It's also interesting as it offers us an example of a transformation process in a culture generally considered to be conservative and deferential to authority. In this context, design acts as a *Cultural Catalyst* and *Friendly Challenger* to challenge the patriarchal and top-down societal norms. At the societal level, design is used to challenge and question the status quo; and at a more personal level, its used to offer honest criticism in a group dynamic situation. The humanising role of design is also used as powerful leverage to motivate and direct participants towards

Who we spoke to
Hiroshi Tamura, Director of Innovation Studio Fukuoka, Director at Re:public
Takeshi Okahashi, Director at Re:public
Yuki Uchida, Director and Researcher at Re:public
Hiroshi Kuboyama, Participant

Why change?

Fukuoka city is the fifth largest city in Japan. Like many other regional cities in developed countries, it has struggled to keep its best talents as many of its citizens often move to larger cities like Tokyo and Osaka for job opportunities. The city council recognises a need to create conditions for ideas and new businesses to flourish. They wanted to encourage an innovative culture that will contribute to the economic and social growth of the city.

Design roles that enabled change in ISF

Types of changes achieved through design

2013-2016

Changing products & services

Changing organisation

What has a design-driven approach brought to ISF?

- Offered a way to exchange honest criticism in a conservative and consensus-driven group dynamics situation.
- Offered a more humanistic lens to counteract the logical and fact driven approach.
- Enabled participants to connect with other citizens/users.
- Offered more creative, visual and open-ended approach to developing idea.

user needs. The focus on the user activates the *Community Builder* role to enable and support the community that is formed around the purpose of doing social good.

Context

Fukuoka is the capital city of Kyushu island, located in the south of Japan. It has a population of 1.5 million in the city and around 4 million in the surrounding area. It is the largest city in Kyushu and has historically been considered Japan's gateway to Asia. It is the only city in Japan that currently has an above average birth rate while everywhere else in Japan has seen a drop. In comparison to Japan's other largest cities, it is able to maintain a high standard of living with comparatively lower cost of living. However despite these benefits, the city (like many other regional cities in Asia) has often struggled to keep its best talents in the city as many seek to move to larger cities like Osaka and Tokyo for job opportunities. The city council recognises a need to create conditions for ideas and new businesses to flourish and also to build on the growing IT community in the region. This will in turn create more job opportunities for the local population reducing the need for local talent to move away.

Citizen ownership through participation

At the core of the Innovation Studio Fukuoka's eco-system is Citizen Ownership. It is about creative values and transforming lifestyles through the use of bottom-up, citizen-led innovation. It leverages the knowledge and experience of informed citizens and couples these with the growing technology community in the city. The platform aims to transform the culture of the city into one that is 'innovation friendly'.

'The platform is not just focused on creating start-ups. We are not limiting the type of innovation to business innovation. It's also about social innovation and changing the way people think of innovation in their everyday life. And that's why its unique and also very challenging.'
Hiroshi Tamura, Director of Innovation Studio Fukuoka, Director at Re:public

The Fukuoka Directive Council was very keen to explore the impact of using a citizen-led approach for a number of reasons. They wanted to encourage an innovative culture that grows what the urban theorist, Richard Florida calls 'the creative class' in the hope that it will attract talent that will contribute to the economic and social growth of the city. They were also aware that many of the issues faced by the city and its citizen cannot be solved simply by relying on company-led innovation, which often can be slow but also

Stages of transformation

1. Planning–Discussion with Fukuoka Directive Council.
2. Implementation–Pilot projects and 3 live projects.
3. Experiencing–Supporting the network beyond the project phase.

INNOVATION STUDIO
FUKUOKA

What can we learn from ISF's story?

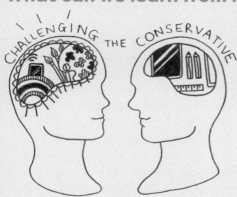

The importance of design as creative tool to challenge and question the status quo, particularly in a culture that is conservative and deferential.

The importance of rallying citizens around user-led social issues to help motivate and direct efforts.

The challenge of maintaining the community once it has been developed beyond a project phase.

counterproductive since companies tend to be secretive and unwilling to share. At the same time, there is a marked change in attitude towards citizen participation after 3/11–which is shorthand used by the Japanese to refer to the March 2011 earthquake and nuclear disaster in Fukushima. This event has proven to be a catalyst in changing the Japanese psyche and has resulted in a rise of bottom-up approaches led by concerned citizens working together to solve community problems rather than relying on support from governmental organisations.

At the heart of the programme is citizen participation and community building. It relies on exploratory design research and prototyping approaches to help participants uncover meaningful insights that are the catalyst for new socially-minded business ideas. It uses a design-led approach that brings out the empathetic qualities needed to view issues and challenges through a more humanistic lens. It also uses co-design methods to enable and encourage community participation and ownership.

'Our biggest challenge is to create an "innovative friendly culture" in Fukuoka. So while we are aiming to maximise the business and economic potential, we are also challenging participants to create and foster innovation with social impact. And that's why we are insisting that it is programme focused, not results focused. It's about using the programme to mobilise, encourage and support an innovative friendly culture to emerge.'
Hiroshi Tamura

The programme
The programme team is led by Fukuoka Directive Council and the Re:public team. The programme is supported by three key sectors: government, industry and academia. It is jointly funded by the city council and local businesses. The programme also emphasises the need for diversity: diversity in the range of citizen participation, diversity in business ideas and diversity in the outcomes. The programme relies on a global alliance, working with different organisations in different countries like Denmark and The Netherlands to learn from and validate what they do.

Since Autumn 2013, one pilot project and three live projects have been completed. Each project is defined by a theme, project one was focused on 'Sports in Daily Life', the second project was focused on 'Life Trajectory' and the third project was 'Uncovering Hidden Resources in the City'. These themes were chosen before the start of the programme in collaboration with the city council to ensure that it covered a range of issues relevant to Fukuoka city. The average number of participants in each project is around 30. The majority of them are citizens and the rest of the participants are made up of industry representatives and specialists. 'Thought Leaders' are experts who have been invited by the programme team to provide mentoring to participants during the duration of the project. They are usually people who have specific expertise relevant to the theme and/or have experience turning ideas into new business ventures.

Uncover, Inspire and Exchange

A project usually lasts for six months and consists of three stages that are recognisably design-inspired. The first and longest phase is the 'Uncover' phase. This phase is focused on helping the participants become citizen researchers. Crucially for innovation, the research phase is not about confirming hypotheses, but about challenging assumptions. This is where design as a *Friendly Challenger* comes into play. The first stage is focused on uncovering insights to inspire new ideas and bring new perspectives to the theme. During the first two months, participants were expected to uncover problems surrounding ordinary people and to uncover new insights that might lead to innovative solutions. Re:public introduced design ethnographic techniques in workshops and helped train participants on how to conduct interviews, make observations and synthesise their findings. Many teams chose to interview extreme users to gain insight into unthought-of behaviour and motivations. Design is used in a *Humaniser* role, bringing participants closer to the citizens' issues and needs.

Teams were brought back together in the second stage 'Inspire', to deliver insights to everyone involved. The two-day workshop involved research insights presentations from all the teams and ideas were generated collectively based on these research insights. Participants had to identify new perspectives, behaviours, and business models for their idea. They also had to identify potential lead users and the type of members needed in order for him/her to realise the idea. Idea owners (an individual who has an idea and wants to take a lead on it) were then asked to pitch their idea at the start of the second day and the aim was to try to 'sell' and attract group members to help them develop their idea. New teams were formed around these ideas and, at this point, the teams had a couple of months to turn these ideas into a realistic and self-sustaining business venture. Re:public ran a number of supporting training sessions during this time to help teams to prototype and communicate their ideas.

The final stage, 'Exchange' was where business ideas were more thoroughly researched, developed and tested to ensure there was a viable route to market. At this point it was important to evaluate how well the idea would be able to scale up and out. Workshops on how to make a video prototype and mapping out a business model using the Business Model Canvas were run with participants. Ideas were presented and evaluated by peers and experts. Potential investors were also invited to attend the pitches to see if any potential partnerships can be formed.

Bringing together a community

Re:public was commissioned by the Fukuoka Directive Council to help design, develop and deliver the engagement process of the programme. Their role was to create a platform that generated new ideas for new ventures rooted in the community. To realise this aim, their role and involvement has been much more complex and far reaching than what a knowledgeable business service consultancy might provide. The platform was essentially a way to bring together and connect a diverse range of participants from citizens, industry, government and academia through shared societal concerns. It offered an opportunity to bring together groups of people who may never have met in their daily lives and leveraged a growing desire to have personal agency to act on issues that have an impact on their collective lives. On the surface the platform may seem like an incubation platform focused on generating new business ideas, but the unseen role of the platform is to bring together and build a community that is focused on transforming society through innovative and self-sustaining ideas.

Its role as a *Community Builder* manifests in a number of ways. For example, 'Thought Leaders' in each project have been a particular benefit to the participants. Not only were the participants able to seek advice and mentoring from the Thought Leaders, they have been able to gain access to their networks as well. Gaining access to these networks has been invaluable and in one example has helped a participant secure the funding required to move the project to the production phase. Thought Leaders were not the only ones opening up and sharing their networks. Re:public has also been very open and generous in introducing participants to personal contacts they have within their network.

'The programme involves specialists and advisors from industry, academia and mentors, as well as attracting people from all walks of life. As a result, the growing network of participants developed through the programme has a really big impact on their lives.'
Yuki Uchida, Director and Researcher at Re:public

What are the conditions for impact in ISF?

- Recruiting criteria: participants have to be open minded but personally motivated citizens.
- Recruiting criteria: Ensuring participation from a diverse range of people (experience, sector, ages, gender and viewpoints).
- The model is suitable for a city with a strong sense and pride of place particularly regional cities.
- A citizen-led innovation platform requires support and funds from local government, business and local community groups.

What have been the challenges so far?

Keeping the community and network going beyond the project phase. Without the formal structure and support from Re:public, it has proven difficult to keep the network intact and the momentum going forward.

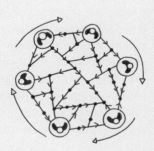

What type of change still needs to be achieved?

- Supporting the new businesses from the initiative.
- Growing and nurturing the community built up from the initiative.

Community building was also focused on connecting local people with local resources. For example, there were collaborations with the local Fab-Lab group during the prototyping training stage. This connection enabled participants to easily create quick prototypes of their ideas and at the same time connect with new people that have the skills and expertise to help them realise their ideas.

Another way in which the *Community Builder* role is evident, is the way the relationship between the participants and Re:public has changed over the course of the programme. Re:public started out in a facilitation role for the participants. Throughout the course of the project, this relationship has organically changed to a *Friendly Challenger* role. The concept of the *Friendly Challenger* (which is often talked about in educational literature as the 'Critical Friend' model) describes a type of relationship that enables existing perspectives to be challenged critically due to the supportive and trusting environment that has been built. This role is somewhat unique to participants because exchanging honest criticism is not very popular or common in Japanese group dynamics.

Further evidence of a burgeoning supportive community can be observed in the participants' interactions and presentations during the workshops. There is a high level of trust between the participants themselves and between the participants and the ISF project team. The atmosphere in the workshops was one of openness, generosity and warmth. At the same time there is a level of criticality and honesty when given feedback, done in the spirit of learning and wanting to collectively improve the idea. It was also obvious that the selection process was important as it enabled the project team to select not only a diverse group of participants representing different aspects of Fukuoka's society, it also enabled them to find participants who were motivated and open to a different way of doing things. On the whole, participants seemed comfortable with the tasks set and the uncertainty of the outcomes. The outspokenness and honesty of some of the comments were also refreshing to see compared with the stereotypical Japanese culture of politeness.

Transformations

One of the explicit aims of the platform is to generate ideas that can be realised as business ideas. Although each project lasted for six months, the programme team acknowledged that it would realistically take more than six months for an idea to be developed and become market ready. They were also aware of the importance in maintaining continued support for ideas and people beyond the six months. There is a natural attrition rate of ideas and teams. On average three ideas survived beyond the initial project duration and since October 2015, three new companies have been established.

Although on paper this does not seem like a huge return on investment, we should recognise how it has had an impact at a more individual level.

What has been evident throughout our analysis of this case study is how the programme has changed the lives of the participants. For many, it has been a personally transformative experience. For example, one participant who was a freelancer, providing consulting advice in human resource development and planning, has now established a new company focused on developing products and services to enhance learning for children. He has become a close associate of the Re:public team and will be launching two new products within a year of participating in the platform. One of the Thought Leaders has become a close project advisor to his business. While another participant from a large telecommunications company has been embolden to set up a new business unit within the company to help develop ideas that emerged from the project. The program has given participants who otherwise were not quite entrepreneurial, the confidence to think of their future businesses from the understanding of society and users.

'One of my key highlights of the project is seeing how participants change throughout the programme. Their ambition builds over time and they are full of confidence and wanting to try new things and seek the next opportunity. They are really motivated to find ways to improve their city, society and themselves.'
Yuki Uchida

The project team has also learnt how important idea ownership is in sustaining interest in the ideas. The concept of identifying an 'idea owner' and a 'supporter' was introduced to support collective ownership of the idea by team members. The flexibility in moving and changing teams is considered by the project team to be a risky thing to do, considering that Japanese people like consistency and stability. However they felt it was a risk worth taking to ensure better ownership and hopefully improve the survival rate of the teams when moving from stage two to stage three of the project.

Roles of design

It's interesting to note that the term 'design' has not been explicitly used to describe the approach used in the platform. However, the process followed (Uncover, Insight, Exchange) is very similar to a design thinking or a service-design approach. This point was acknowledged by the Re:public team. It might have been easier for Re:public to refer to the term 'design thinking' since the term has become quite popular in Japan in recent years. However, the Re:public team was also wary of the baggage that the term brings and wanted to go beyond what they perceive to be its limited framework. Despite their reservations regarding how narrowly design thinking has been understood and operationalised, the team was still conscious that they were using design-inspired approaches and tools to help them foster a citizen-led innovation culture.

Design has offered the project partners an approach that enables them to focus and leverage the humanistic aspects of society. This approach challenges the dominant logical approach and also offers a powerful counterpoint to it. As mentioned earlier, while the explicit aim was to create new businesses to generate economic growth, the longer term and more transformative aim was to inculcate a culture of innovation and creativity in the citizens through participation. Focusing on the emotional core as the anchor and driver of change has been a key strategy to build and hopefully maintain a culture of innovation. Participants recognise the value of user research and tools; a customer journey map was cited as an important tool to help them understand how a person experiences a service in their context.

'Many of our participants joined the programme because they were either in a life-changing situation or wanted to change the direction of their life. We are challenging not just their rational mindset, but trying to involve their entire being, tapping into their emotional side. It's really about the ownership of their life and we believe that the design approach helps them connect with their humanity and also helps them connect with others in a more humane way.'
Yuki Uchida

Design research does not set out to prove a hypothesis but rather to uncover hidden insights, challenge assumptions and to highlight new opportunities. Participants were asked to consider the new behaviour, methods and values that they wanted to evoke in their lead users during idea generation. This 'Future Visioning' process has helped give participants a focus that is not just about revenue generation but also about creating a better life for their users. Design tools have also been helpful in changing the way people respond to certain types of situations. For example, when being asked to come up with ideas, participants were encouraged to find more visual and playful ways to communicate their idea. Working visually and by 'learning through doing' have opened up new creative spaces for the participants.

'We use design research methods because they are very much focused on revealing and understanding the end user. It highlights the pain points of the customer and what core values they embrace. This helps us see the potential customers and to organically help us take ownership of their problems and values as well.'
Hiroshi Tamura

'Sometimes the participants' research findings are not so cutting edge, but they really benefit from having gone through the process so they are able to change they way they view an issue or approach a problem.'
Takeshi Okahashi, Director at Re:public

It's clear that one of the key roles played by Re:public and the Innovation Studio Fukuoka platform is one of *Community Builder*. It has enabled a community of like-minded people to meet and form informal networks and help build a socially-focused vision of Fukuoka city. While it remains to be seen how the initiative goes forward beyond its current funding period, it's clear that design has an important role to play in catalysing a culture of innovation. It has brought communities together and humanised the process of innovation.

Section 2
Expert interviews

Peter Coughlan
Consultant, US

GK VanPatter
Humantific, US

Mark Vernooj
THNK, The Netherlands

Mariana Amatullo
Designmatters, US

Brenton Caffin
Nesta, UK

Christian Bason
Danish Design Centre,
Denmark

Beatriz Lara Bartolomé
Imersivo, Spain

Organisational change through design

Peter Coughlan

Peter Coughlan is an organisational design and change consultant. He has over 20 years of experience working with a variety of organisations in strategy, innovation and design. While at IDEO (a world leading innovation and design company), Peter established and led their Transformation by Design practice, helping client organisations such as Kraft Foods, eBay, Eli Lilly, Hewlett Packard, McDonalds, Memorial Sloan Kettering Cancer Center, the NHS, Proctor and Gamble, Steelcase, and Turner Broadcasting, to imagine and implement futures of their own design.

What is a change consultant and how did you end up being one?
My role as a change consultant is to help clients envision and implement new futures for their organisations. This can result in new products or services or a change in the way the organisation is run. I started my career by learning to apply ethnographic research to design whist working in multidisciplinary teams at the Doblin Group (www.doblin.com) in Chicago. At IDEO I moved to an organisational change role because we saw how organisations struggled to implement new ideas and solutions throughout the innovation pipeline. We realised that it was really about building internal empathy and resistant teams.

Between working at the Doblin Group and IDEO, I finished graduate school and worked at Nissan for a short while. My experience at Nissan showed me how important user needs are. Nissan was just beginning to embrace a user focus at that time so I had to find ways to demonstrate how user needs could inspire and inform our designs. I realised the importance of involving people very early on in my career.

At IDEO, even though the ideas we came up with were often quite compelling, it was sometimes difficult for our clients to implement them. I began to realise that generally the problem of implementation was related to our client's existing organisational structure rather than the quality of the ideas. And so from that early work I helped develop a group within IDEO that started to focus on helping organisations with change using a design-based process. We started the Transformation by Design group in 1999. Our premise was that all change should stem from some human need that is not being met. Our approach was to observe how our clients were working, managing and communicating so we began to uncover the gaps between what they were currently doing and what they could be doing to achieve the end results they desired.

We started playing with the notion of organisational prototyping and moving it towards what we would later call 'experiments' which is similar to the approach and process of the LEAN start-up model. It was real world, in-context feedback as opposed to controlled, closed-environment feedback through means of focus groups and the like. These experiments helped our clients to try out new products and services in the field and to collect meaningful feedback. It was also a useful way for the idea to gain traction in the organisation, not just from the employees but also from the actual users. Having this direct interface with the customers helped them overcome the organisation inertia they had.

So in summary, I became an organisational change consultant because I realised that in order for innovative products and services to be successful, we need to focus on how organisations are structured to create and deliver these products and services. Design provides a mindset that can help organisations understand and embrace the change needed to bring new things into the world.

What role does design play in helping organisations change and what is its key contribution?

Creation of experiences

For me, design is, in its most fundamental sense, the creation of experiences, whether they are through products, services, spaces or organisational structures. So it's really important for innovation because it helps reveal either gaps in the current experience that is being designed or it helps reveal experiences that could create additional value for new and existing customers. The most valuable part for me is using design to help an organisation frame what it does and this will then enable them to see the opportunities beyond the experiences that they are currently creating. Innovation is the 'why' but design is the 'how'.

Humanising the organisation

Going out into the field collecting discrete data points by observing how people behave is a really valuable design habit. It is incredibly valuable because you take these discreet data points and use them to identify opportunity areas. You can also use them to identify principles that can then be used to generate and evaluate ideas. The process also helps clients visualise experiences, and typically at a much broader scale than what they had been thinking about. So tools like 'a day-in-the-life of' or a 'customer journey map' are incredibly powerful for clients. Prototyping also helps organisations create intimacy with customers by connecting and understanding their needs.

Bringing different stakeholders together to envision possible futures is really important since it gives humans a sense of agency over their lives. Enabling the whole system to be in control of its future by designing it together is a really powerful concept. And then, of course, encouraging people to experiment and prototype, not just at the level of the offer but also at the organisational and system level. By involving multiple stakeholders, we are able to observe how this new experience unfolds and affects them in the system as well as reveal all the unintended consequences that have been co-created. This means that it then becomes more difficult to blame people when things go wrong. It shifts the process from one that is fear-based to one that is inquiry-based.

For the last twenty years you have been involved in helping people in organisations implement and manage change. What have been the significant changes in this practice and why?

Increased design literacy

First, awareness levels and understanding of design have changed. I work closely with clients to help them identify opportunities and provide support throughout the process up till delivery. When clients are able to apply the same process to address a different challenge, it indicates that they have understood the value of the process and know when to use it. I see this happening more often now, as it's more common to use design to help organisations change.

Related to this, an increasing number of organisations are establishing in-house design teams. As a result, the overall level of design literacy in organisations has risen in the last twenty years. Now it's not about arguing for the value of design but arguing if they are doing it well enough or if there is an opportunity to improve the way they are doing design.

A third way

There is another interesting cultural shift happening in the larger society. Harold Nelson and Erik Stolterman speak about this in their book, The Design Way. They talk about design being this third way of thinking, somewhere in between the way that artists think and the way that scientists/engineers think. Design is this beautiful melding of the two. I see design becoming more important because the world's challenges are getting to a point where we now understand that they can't be solved with the standard science/engineering mindset, nor can they be solved with a purely artistic creative mindset. There needs to be some blending of those two modes of thinking. There are a growing number of people in Silicon Valley that consider themselves to be designers. These individuals are comfortable working across the spectrum, from writing code to designing beautiful interfaces. This is a really interesting shift.

Empathy with one another

Another key shift is that we are evolving from using empathy to reveal customer needs to developing empathy for all stakeholders within a system. And so in my own work in particular I'm using some interesting methodologies like Kegan and Lahey's Immunity to Change model. I'll typically start a design process by having conversations to uncover barriers that are stopping people delivering the expected customer experience. For example, 'What are the competing commitments in our organisation?'; 'What are the hidden assumptions or hidden drivers that prevent us from coordinating or delivering an experience?' or 'What are the structures in place that cause different factions within the organisation to compete with one another?'.

In the past, design's focus was more contained and specific, for example designing a product or a great experience. I feel that design has matured to a level where it is capable of operating at a systems level. Of course, that itself is a massively difficult thing to do. We as designers need to learn skills that help us see in systems and to uncover hidden structures that shape these systems. Then we need to develop additional competency in helping facilitate groups of diverse stakeholders to do the same, because ultimately, human systems tend to reject that which they haven't themselves created. It goes back to something I said earlier–the best designs in the world will not see the light of day if we haven't taken into account all the visible and invisible structures and dynamics at play in a system. That can best be done by helping the system see itself and learn to design responses to what it sees.

Self-awareness and continuous learning

Organisations that are going to be successful at delivering great experiences have to be self-aware as organisations. Not only do they need to see their products and services as designed experiences, they need to see the organisation as one as well. Organisations need to be able to hold and address these two concepts at the same time. They need to understand and consider their business ecosystem, which includes their partners, competitors and other stakeholders. So if you can develop empathy for all the stakeholders in your ecosystem, you will be able to design and deliver better experiences as a whole.

Some organisations are starting to recognise that, to remain successful, you must embrace some form of continuous learning process (as Peter Senge started telling us decades ago). Successful organisations are opening their boundaries, and inviting customers, stakeholders and even competitors to co-learn. Design is a form of learning. It's about putting out new ideas through products and services into the world, seeing how they impact peoples' experiences, and then iterating based on what's been learned.

What have been the key challenges in designing, implementing and managing change through design?

Design as a process

In some ways, the design process is very easy to learn, at least at a basic level. So clients can be lured into thinking that they have transformed the way they think and approach challenges by simply following a process or using lots of post-it notes. The trick is to help clients learn enough so that it begins to disrupt their usual way of thinking. Then, the next time they're faced with a challenge, they think "hey, I know how I usually approach this challenge, but wouldn't it be good to look at it from a fresh perspective. How are other industries solving this problem? How do some of our extreme customers view our experience?"

Multi-disciplinary skillsets

It is still rare to find people with the right mix of skills to help organisations change. They need to be able to draw on a range of tools and expertise from design, to systems thinking, to organisational change, and be able to use a range of different set of methodologies–from observation to visualisation to conversations around Immunity to Change and facilitation structures like World Cafe. Not least, they also need to be good designers. They may not need to be real craftsman but they need to be conceptually strong as designers. Even though the field of design has grown a lot over the last twenty years, it hasn't grown to the point where every organisation has great designers, great systems thinkers and organisationally savvy people. Since design schools do not typically teach these skills, it will be a challenge to find enough people capable of cobbling together an education that spans these disciplines.

Timeframe

Most organisations are still unwilling or unable to invest in the three to five years required to really make an organisational shift of that magnitude, and that's on top of the personnel problem or the talent problem.

Creating protected innovation spaces

Experimentation is really difficult to do. Even though prototyping and experimentation are meant to reduce risk for the incumbent organisation, it challenges the existing structures. Vijay Govindarajan and Chris Trimble, in their book 'The Other Side of Innovation' identify the tension between the

'Performance Engine' team and the 'Dedicated Team.' The former relates to day-to-day operations while the latter is focused on innovation. They talk about the inherent conflict between the two teams due to the risks taken by the Dedicated Team that could potentially be disruptive to the structured process of the day-to-day team. So increasingly I'm trying to get my clients to see that they may need to create completely different structures in order to make their new-to-the-world stuff happen.

I used to be really against the Skunkworks approach–a small team dedicated to working on radical or disruptive innovation outside of the normal organisational structure. However, I have adapted it in some organisations so that ideas go into a protected space after they've been developed in the organisation rather than before. This reversal offers everyone in the organisation an opportunity to contribute, but also recognises that current organisational structures are not conducive to more radical or disruptive innovation. By doing this, it creates a firewall between the idea and the need for it to grow and flourish outside of the current structure. Another thing that I'm seeing organisations do more and more to satisfy their need for innovation is to acquire or invest in start-ups that have developed ideas in this space. I predict that in the not too distant future this will become a normal part of most large organisations' strategies.

What does it take for an organisation to be ready for that change?
Long-term commitment
Organisations need a long-term mentality. Sometimes it's difficult if the incumbent may have already tried different methodologies to invoke change. One of my recent projects involves working with someone who has just joined an organisation in a leadership position. He recognises that turning a big ship around will take time, so he has made a long-term commitment to the change process and has set more modest short-term goals to start.

A vast majority of organisations still promote people on a yearly to a year-and-a-half basis. Therefore in light of the longer-term investment needed you may lose people that were initially excited about an idea. They might then be deployed elsewhere in the organisation. That can be a good thing, but it can also mean that you're in this continuous state of restarting. My solution to that is to let people take ownership of what they're innovating and encourage it to grow as part of their work.

Start small
The person leading the change needs to be persistent; and, paradoxically, they need to be willing to start small. This is how I work with my own clients now. I start very small, very humbly. I try to keep the communications people at bay since I prefer to see if it works first and to learn from it. Eventually when

word gets out, the work should hopefully speak for itself rather than through a carefully crafted communication that's designed to amplify impact before true impact has been achieved.

Embedding design principles

Having a conducive organisational structure is also important. Newer organisations tend to have flatter profiles and build principles that embody design thinking into their charters, definition, rules, structures and ways of working. If design is embedded from an organisational structure point of view, then it's not just dependent on a person or the availability of funds.

It's also important for an organisation to recognise the universality of the application of design. It's about being exposed to the different stages of the process and also about bringing in people from the outside to these teams. Once someone has participated and experienced the process, they are more than likely to champion the use of design in their own work.

Sensemaking to Change-make
GK VanPatter

GK VanPatter is CoFounder of Humantific and an internationally recognised innovation strategist, methodologist and capacity building advisor. His long-standing work includes the integration of visual sensemaking into the innovation process. He has more then a decade of experience working with organisational leaders on designing, driving and explaining organisational change. Former editor of NextD Journal he writes frequently on the subjects of strategic design thinking, inclusive innovation and the future of design that has already arrived. He speaks frequently at conferences on subjects related to innovation enabling in organisations. Humantific's new book entitled Innovation Methods Mapping: De-Mystifying 80+ Years of Innovation Process Design will soon be published.

Can you start by telling us a little about what Humantific does and the focus of 'SenseMaking' to help organisations develop and build their change making capability?

Hello and thanks for your interest in Humantific. Since 2002 we have been working with organisational leaders seeking to create, drive and explain change in their organisations.

At the foundation of what Humantific does is the realisation that Elizabeth Pastor and I had early in 2000 that there is a not very well understood connection between information and innovation. Our prior work with Richard Saul Wurman in part informed this realisation. Since that time we have been engaged continuously in better understanding the connection between what we now know as sensemaking and changemaking by working in the real world with organisational leaders.

Today enabling interconnected sensemaking and changemaking provides us with a rich landscape of possibilities for how we help our clients who seek to tackle complex, fuzzy challenges or build next generation innovation capacity in organisations.

What makes Humantific different from many of the firms that appeared in your earlier book 'Design Transitions' is that methodologically we don't make any assumptions upfront regarding what the organisation's challenges and opportunities are.

In the marketplace we see many 'Design Thinking' firms engaging with situational product/service/experience design methods that inherently presume upfront that the challenges and outcomes will be product/service/experiences. In the work that we do Humantific makes no such presumptions in our methodology.

Since we have been working in this arena for a long time we already know that organisations and societies typically are facing a vast array of challenges, not all of which can be assumed to be product, service and experience related. Once we acknowledge that relatively simple notion, many of the traditional industry methods and offerings of Design 1 & 2 fall into more precise context.

We already know that situational methods are most useful downstream. With their assumptions baked in, forcing the application of downstream methods upstream tends to create poor, biased challenge framing and considerable confusion. We see this methodology confusion in many organisations today.

How is sensemaking linked to design?

To briefly explain how sensemaking is linked to design we often utilise a construct created by JJ Gordon of Synectics. In the 1960s Gordon made the useful distinction between what he called 'making the strange familiar' and 'making the familiar strange'. Today we recognise that all design projects contain these two elements in varying proportions to each other. The former we call sensemaking and the later strangemaking or changemaking.

Most design school portfolios that we see in candidate interviews today are filled with object-oriented strangemaking projects. This is what the design schools have been teaching as design for decades. The branding industry is all about strangemaking, how to make one toothpaste or bottle of water look different from another. Object oriented differencing has a long history in design. Differencing is a specific value-add that remains an important part of design.

At Humantific we recognise that sensemaking and strangemaking are two quite different mental operations requiring different skills and toolsets. Both add value.

Around 2005 we mapped the notion of sensemaking and strangemaking to the NextD Geographies framework depicting the differences in methodology across Design 1,2,3,4. This helped us figure out the relatively simple notion that the proportion of sensemaking and strangemaking alters, inverts as challenges scale. The more complex the fuzzy challenge, the more sensemaking is required.

The simple NextD Geographies model has, from our perspective, tremendous and unavoidable implications for design practice and for design education. Globalisation has changed design practice in ways that are difficult to argue with. We have been writing on this subject for a decade.

Our book is premised on the idea that design creates value, and specifically we are focussing on how it helps organisations innovate and transform. In your opinion, what role does design play in this context and what is its key contribution?

Adaptability remains one of the most enduring goals in organisational readiness and transformation. As a need and a goal adaptability has endured through the ages across many generations.

Charles Darwin is credited with famously saying: 'It is not the strongest of the species that survives, nor the most intelligent that survives. It is the one that is most adaptable to change.'

Certainly the business, organisational leaders working with Humantific recognise that static entities tend not to survive. As a goal adaptability keeps getting creatively repackaged by each generation. As a business need it has certainly increased in importance in a now continuously changing world.

To put this in popular business media context: A few cycles ago Fast Company published, with considerable fanfare, an issue heralding in what it depicted as the arrival of 'Generation Flux'. Take a wild guess what that was all about.

Readers who were aware of innovation dynamics history would recognise that adaptability, flexibility, agility, 'fluxabilty' are all different ways of saying more or less the same thing and that thing as a capacity for organisations has been a recognised need in American business organisations since the 1950s. It is how adaptability/fluxability gets enabled that has evolved and changed significantly.

Adaptability and efficiency are recognised as two very different things. Efficiency is about doing the same thing better. Adaptability or agility or 'fluxability' is about continuously, proactively identifying and actionising how the organisation needs to change...and changing it.

In terms of innovation and transformation, there is today an added wrinkle in play that increases the complexity of marketplace relevance. Today many organisational leaders want to do both: make the most of what they presently do while simultaneously creating new paths and possibilities. Many leaders have come to the realisation that one or the other is no longer enough.

In the business management literature this dual engine strategy has been framed as ambidexterity as in Ambidextrous Organisations. In that stream of literature the two dimensions are often described as Exploration and Exploitation. Currently the CEO community has considerable interest in enabling this dual engine strategy. This is essentially where Humantific operates.

What Humantific does is bring the Ambidextrous Organisational strategy to life as human-centred, inclusive innovation. The three-step dance shift underway involves beginning upstream instead of downstream, undertaking sensemaking and changemaking, while realising exploration and exploitation.

Everything we do syncs with a visualised ambidextrous model of innovation, rather then the more traditional single engine model. It is true that to realise that ambidexterity model we make use of tools, behaviours and dynamics from design as well as from other discipline expertise that all interconnect with ambidexterity in one way or another. It is literally how Humantific redefines human-centred innovation today. Enabling ambidexterity is how we get to adaptive human-centred organisations.

You have pointed out that lot of the innovation methodology ideas, concepts and tools have been developed by the Applied Creative Community. However they never really caught on in the business community. Why do you think this is so and why is design managing to make the leap across?

The arena of business change and innovation is vast and has been around for decades, involving many players not just design firms. Not everyone in the design community seems to be aware of this but most leading strategic design practices operate with hybrid toolboxes, many components of which originated in the applied creativity community of practice.

That's a community that predates the design methods movement and has been around since the late 1940s and 1950s. It's a community that has deep methodology knowledge that relates in particular to organisational contexts. Lets acknowledge that organisational changemaking has not been the traditional role of traditional design practices. In fact still today many design firms have no interest in becoming involved in organisational change or culture change related work.

Interest in various terminologies tends to ebb and flow as the media looks for fresh topics and often in parallel, communities look for ways to reinvent their own practices and learning systems.

The truth is, interest in design thinking arose as the business education community finally recognised the long overdue need for change in that community in the context of the innovation era in order to better reflect organisational needs. It was a rather late awakening but better late then never! Understanding design at a rather primitive level they grabbed design thinking as a device to transform their own knowledge and ultimately their own offerings. Often what they grasped onto originated in the applied creativity community but those subtleties are often lost in an over-simplified marketplace.

The boom of interest in the wrapper 'design thinking' has been a mixed blessing as it has heightened general awareness in the business community but it has also dampened down recognition of the need for change within the design education community. The massive interest in design thinking has served to block much needed forward change in design education, which to an unfortunate degree continues to re-spin old programs and methods. The spin coming from the mainstream graduate design schools has created a mountain of confusion around design thinking in the marketplace.

This kind of chaos and free-for-all makes for not necessarily high quality but certainly lots of choice in the marketplace. In the messy context that now exists there is certainly a need for community sensemaking. We have been writing into that space for many years. Books such as this one that you are preparing will hopefully contribute to more clarity.

For the last 15 years you have been involved in helping people and organisations implement and manage change. What have been the significant changes you have observed in this practice and why?
We see a lot of good news in the various mega trends underway in the global marketplace that ultimately impact the industry. Those shifts include:

- increasing CEO interest in navigating complexity
- linking innovation strategy more closely to corporate strategy
- enhancing the performance of cross-disciplinary teams
- organisational sensemaking related to so-called big data
- sensemaking beyond big data
- integrating 'evidence' as well as hunches into innovation process
- building advanced 'problem finding/solving' capacity
- enabling multiple participant co-creation
- embedding human-centred innovation skills in organisations
- skilling-up organisations to better address internal and external challenges emerging in a continuously changing world

All of those trends have already changed the nature of strategic design in practice. Most of those trends dovetail with what Humantific has been doing since our founding so we are happy to embrace the arrival of such shifts even if we often have a different point of view than mainstream media.

For numerous years Humantific was far out in front of the marketplace. Today there is growing awareness in the business community regarding what methodologies work best in what organisational contexts. That's all good news as far as we are concerned.

The strategic design practice community is not an abstract business idea. Its existence is being driven in large measure as a response to globalisation and other shifts. With those shifts in mind the future of design thinking is likely to involve helping others tackle more and more complex situations reflecting, not the mantra of design, but rather the real world in all of its fuzzy complexity.

What have been the key challenges in designing, implementing and managing change through design?
Here are three often-appearing challenges:

Challenge finding acceptance
Challenges are identified but not everyone likes them. This dynamic is seen particularly in large organisations and in societal problem-finding contexts.

Challenge ownership
Organisational leaders have to own their challenges and related initiatives otherwise nothing much happens.

Behaviour change
It is not enough just to have innovation stated as a value on a wall plaque hung at corporate headquarters or to assume that technology makes everyone instantly, magically innovative. Behavioralising innovation so it becomes real in everyday operations remains the heavy lift that takes real commitment.

How long does it take for a transformation to occur in a company and what are the different stages to them?
To get at this intervention terrain we must first move away from the traditional design notion that we are working from one brief. Think 'challenge constellations' not a single tactical brief.

The transformation business involves helping organisations surface and organise what their real challenges actually are and as part of the process tackling all kinds of challenges. Typically there is no one 'transformation'.

Change related projects take many forms. Big picture strategic challenges such as: 'How might we become an industry leader?' or 'How might we become a customer-centred organisation?' typically have hundreds of related sub-challenges beneath them that all require attention.

Cocreating big picture maps of the 'Today' challenge constellations, with multiple constituents participating, is among the first cycles and many organisations need help to do that effectively.

Once the map is realised the organisation has a logical plan of attack that typically involves multiple simultaneous streams of activity, evolving through stages of change that map to the innovation process.

Every organisation is different. Each transformation attack plan is different. How to get from today to tomorrow is essentially what the plans are all about.

We might start from today and move forward or start from tomorrow and move backwards. There are several options in terms of how to get to the tomorrow picture and all of them involve co-creation. Without co-creation by multiple participants transformation often stalls with roadblocks. In Humantificland facilitation is a process role not a content role. We are not there telling the organisation what to do. This is often a huge shift in logic for folks coming from (MBA) business and or (MDA) design backgrounds.

What are the optimal conditions for design to be effective in this context?
We think of what we do as enabling innovation, not design. For us this includes business, non-profit, government and community organisations. The optimal conditions are the existence of problematics coupled with the recognition of the need for change. To be most effective organisational leaders have to skill-up and own-up to their change initiatives.

Perhaps most important in this question is to recognise that the design we are talking about, in terms of what it is and does, might not be the design that some of your readers have in mind. I am not referring to the downstream methods or tools of Design 1 or Design 2 here. If we don't point that out readers might be left with the wrong impression.

A rough analogy would be that we are constantly engaged in reformulating the 'design airplane' and its fuel while it is in flight into new terrain. When we are referring to design or design thinking, we are referring to these new constructions, under construction, not the old constructions of design.

Of course there is often a lot of promotional talk in the design community regarding working with 'leading companies' but most often it's the organisations who are not leading that need help.

Typically today we are not there to introduce the importance of innovation or design thinking. Organisational leaders around the world already know this mindshift part. What leaders are seeking our help with is the 'how to make cross-disciplinary innovation real' part, what we call the skill-shift and culture-shift parts.

This is where Humantific most often helps. We bring clarity, practical enabling strategies, clear explanations, tools and inclusive methods that can be adopted by all disciplines. We can explain this in the context of design thinking or innovation whichever makes most sense to the client.

Connected directly to stated visions, values and strategies this form of innovation enabling is difficult to argue with. To be effective you have to make the case relatively bulletproof. If you don't, nothing much is going to happen.

How do you ensure change is implemented and managed appropriately? And how do you solve the problem of legacy and scale?
It might be useful at this juncture to point out that we don't operate like a business consulting practice. We are opposite to McKinsey. We don't send a huge team to camp out for a year and we are not there telling organisational leaders what to do in terms of content.

The practical aspects of 'ensuring' are inevitably connected to limited project duration so we have to assume up-front that the initiative owners take responsibility for their initiative. Inevitably at a certain stage we move to an intermittent role and thus the importance of transferring process skills.

Teaching transferrable process skill is also how we deal with scale. No longer is there a special group of innovators or 'creatives'.

The legacy that we most often encounter that is negatively impacting innovation is the privileging of convergent thinking (decision-making) and we deconstruct that legacy in the context of enabling ambidexterity. We do that by doing brain surgery on the legacy, redesigning privileging to better reflect the organisations stated values and strategies necessary in the innovation era.

The practice of using design thinking to help organisations innovate and transform has been growing significantly over the last 10 years, not just in the western developed world, but significantly in the developing countries. What will be the next shift in the practice?

Its no secret that the design community exists as an amorphous time warp, meaning that parts of it are always in different places along a development time line simultaneously. There is no one present state so there is no one future state. If you think across the activity of Design 1,2,3,4, some of those circles are shrinking and others are expanding. In part that modulation is the future of design so there is lots of choice for practitioners there.

The future state stream that we began talking about ten years ago recognises that strategic design is on track to become the most active and growing circle of activity in the multidimensional future of design.

In that future we expect more emphasis on clarity of methods. We expect there to be more emphasis on better understanding and preparing the fuel for innovation as part of the innovation process.

And finally-given the subject of our book, what is the one key point that you will want to take away from the book and why?

In order to remain relevant to organisational leaders we have to step away from the mantra of design to understand what is now required. Meeting the complex needs of organisational leaders requires fundamental changes to what design is and does.

Creative leadership and design thinking
Mark Vernooj

Mark Vernooj is Partner at THNK School of Creative Leadership (www.thnk.org). He is passionate about innovation, entrepreneurship, education and solving difficult problems. He uses his entrepreneurial background and his experience as an innovation and strategy consultant (with Accenture and McKinsey) to help individuals, start-ups and large organisations become more innovative. At THNK his focus is on developing and delivering their executive program, in-company innovation programs and online innovation (online courses, platforms and tools). Mark shares his view on creative leadership and the importance of knowing when to use design thinking in an organisational context.

What is creative leadership?
Creative leadership for THNK is all about helping people to develop new ideas, supporting them to make those ideas happen, and then showing leadership in that context. Ideas have to go hand-in-hand with making things happen. Creative leadership is not about inventing new things but more about designing new business units and movements (for example in social entrepreneurship). There are prime elements to creative leadership. We help people figure out what they're good at, what they are weaker at, and then we typically help them become better at what they're already good at rather than fixing their weaknesses.

What is THNK?
Model
What and how we teach at THNK is very different from other schools. We are not an MBA for creative people. Nor are we a creative school for people that already know business. We are neither a business school nor a design school. One could describe THNK as an accelerator, a school, a start-up, and as a consulting company. There are only a few organisations like us. For example Kaos Pilot in Denmark caters to a younger audience just out of high school while we run an executive programme. We came from the innovation/creative angle and added leadership to that. Kaos Pilot comes from the leadership angle and added innovation to that mix.

Programmes

We deliver three types of programme. Our flagship programme is the 'Executive Programme'. It's a very intense programme. People enrol in the programme for either six months or a year and a half, depending on the location. They meet four times in the first six months for a week. They receive coaching and mentoring between those meetings.

The second type of programme is our 'In-company Programmes' and they are programmes that we run with a range of organisations: corporations (e.g. Shell, ING, Facebook), government ministries (e.g. Dutch Ministry of Defence), city councils, festivals (e.g. Burning Man), world governing organisations (e.g. FIFA) and non-profit organisations (e.g. World Economic Forum). This type of programme is highly tailored. It leverages all the expertise, research and curriculum that we have. For example we might repackage a 40-day curriculum into a ten-day intense course tailored to suit the individual organisation.

The third type of programme that we offer is 'Open Enrolment Programmes'. This is the type of programme we are running at Stanford University and are hoping to be able to offer it in different locations around the world. Additionally, we are currently developing a digital platform called Collaborne that supports companies aiming to drive innovation inside their organisation. The aim of the platform is to guide people with ideas through a design process. There is a strong collaborative element where peers can offer feedback and support people to turn ideas into insights; identify user needs and test through prototyping. It also becomes a repository of ideas built up over time in an organisation.

What are the characteristics you look for in your applicants to the Executive Programme?

Formal criteria

We have two formal criteria. We generally don't accept anyone under thirty because when you consider the characteristics of a leader, we would expect that a person have reflection and introspection skills. These skills typically emerge a little later in life, after that person has gained more life experience. Our second criterion focuses on a continuous progress in their career. We want to work with people who are on the rise, who are moving upwards and are already at a phase where we're scared of them. It's something we like.

Characteristics

Then there are two characteristics that we look for in a person. We look for what we would call the 'explorative mindset' and someone who is passionate and purposeful. Someone with an 'explorative mindset' is always curious and wanting to learn more. You sometimes see it from their CV when they jump between different jobs. So, that's one important characteristic. We don't have time in forty days to nurture this mindset. It's the same case for passion and purpose. We want to work with an individual who is really excited about getting something done, has true passion and a fire from within.

What is your view on design thinking and where does it fit into the idea of a creative leader?

I talked about three areas in relation to creative leadership: coming up with great ideas, making them happen and showing leadership in this context. Design thinking is an amazing tool to generate great ideas for very specific problems.

Design thinking as I learnt it and as it's been propagated by the likes of IDEO and d.school can lack a strategic perspective in relation to problem definition. What is the problem we really need to solve? Where are the biggest levers? Are we solving the right problem? It's also difficult to scale solutions identified through a design thinking process. Scaling, for us, is already relevant in the design phase, not just the implementation stage. In Scaling we ask ourselves the design question: how would we design a product, service or movement in a way that it spreads and grows? Additionally the process doesn't just end at the prototyping stage, it's about how it is implemented. A typical design thinking process doesn't really take this issue into account. So while design thinking is useful for the idea generation, it's not specifically designed for implementation and developing leadership.

What challenges do organisations face when they're trying to innovate?
Selecting the right tools
There are a couple of challenges that companies face. One of the more interesting challenges is something I call, 'selecting the right hammer'. For example, I don't think many organisations know when 'not' to use design thinking. The scale and context of some problems are just not suitable for a design thinking process. If we use the process of building the biggest ship in the world as an example, you can't just decide halfway through the project to change its colour to yellow. The idea of quick iteration cycles does not always work in large, capital-intensive projects with a long lifespan.

Another challenge is about finding the most appropriate language to guide and describe the activity. There are many different processes and conventions but having that language and being able to say 'we are now sensing' or 'we are now ideating', or 'we are now prototyping' is very useful.

When should you use design thinking?
There are two frameworks I have been using to help me answer this question. They're not flawless but they are useful.

Known/unknowns

The first considers how concrete or abstract the question is. How much is known and unknown about the question. So for example, if there is no clear user-relevance, then I typically don't think it's a design problem. Similarly if it's a technology-focused issue (e.g. making something smaller or faster), it is probably too concrete as a design thinking problem. If you know what the answer is then typically this is not an innovation question, not a design thinking question and not a question that we would answer through a design thinking process.

Industry

Another way to look at this is by looking at the type of problem at an industry level. This lets you look at a strategic level rather than a problem or operational level. If you look at the problem in this way, you can plot the clarity of the industry future (clear or not clear) versus the typical investment size (high or low); this gives you four types of industries with four typical innovation approaches. This then will help you anticipate how and when to use design thinking.

To summarise, what would you say makes an ideal creative leader?
There isn't just one ideal model. The obsession about becoming the next Steve Jobs is nonsense. That typical profile only worked for a very specific industry, situation and company culture. It's highly likely that in all other situations it would have totally failed. And even in that situation it failed most of the time. Many of his ideas never made it to market and not all Apple products were a success. What is important is:

Awareness of strengths and weaknesses

So there are many models of a creative leader. I admire different creative leaders for very different traits. A common denominator that I have observed in them is that they are really clear about what they're good at and are also very good at surrounding themselves with people to help them with things that they are weak on.

Passion, purpose and an explorative mindset

Passion, purpose–a clear and compelling reason to be doing what you're doing and an explorative mindset, a high level of curiosity–are critical. These components are both essential and very hard to build; that's why we use them as selection criteria for our programs.

Aggressiveness

Finally, all creative leaders that I admire have some form of aggressiveness in their thinking and acting and very often it seems to come from insecurity or from having the feeling of needing to prove oneself. That aggressiveness often translates and manifests differently in people. For some people, it's about thinking big-aiming to redesign the world rather than just a new water bottle. And in others the aggressiveness manifests in the way they make deals and negotiate. Others are very aggressive in getting all the small details completely right-like how some architects are known to be obsessive about details. Creative leaders use that aggressiveness or drive to help them move forward and get things done.

Organisational design in a social innovation context

Mariana Amatullo

Dr Mariana Amatullo is the Co-Founder and Vice President of Designmatters at ArtCenter College of Design, Pasadena, where she has overseen an award-winning portfolio of educational and research collaborations in art and design education and social innovation since 2001. She is the lead editor of LEAP Dialogues: Career Pathways in Design for Social Innovation (DAP, 2016) a publication that brings together 84 design leaders shaping the field of design for social innovation in the United States. She received her PhD in Management from the Weatherhead School of Management, Case Western Reserve University in 2015. She is presently a Scholar-in-Residence at the Weatherhead School where her research focuses on the impact of design in social innovation and organisational practice.

Can you start by telling us a little about what you do? How did you find yourself working with design?

I encountered the field of design and design education when I joined ArtCenter College of Design in Pasadena, after a few years trying my hand at curatorial work in Los Angeles art museums. I was quite fortunate then to be given an expansive mandate: to investigate opportunities for innovative educational programming that would open the college up even further to the world.[1] The outcome was Designmatters, a college-wide program dedicated to the intersection of art and design education with social innovation through project-based, experiential curricula. In fifteen years of practice, Designmatters has evolved to become ArtCenter's social impact educational department. In my current role, one key component of what I do is to help conceptualise social briefs and lend strategic support to our faculty who are at the helm of an immersive set of courses and multi-disciplinary projects. Our aspiration is to offer our students a unique set of learning outcomes that come from the constraints of designing at multifaceted levels of complexity and community. I am also dedicated to building with my team the necessary ecosystem of conditions with partner organisations that maximise the chance for project outcomes to be viable and make a positive impact.

Our book is premised on the idea that design creates value, and specifically we are focusing on how it helps organisations innovate. Looking at the seven roles we have identified so far, which do you think most resonate with your experience and the work that you have done?
The seven roles you have identified speak of design's potency at two key levels in organisations: decision-making and what I would call 'insight-making'. We can view the former as more concrete, operational and tangible: i.e. your reference to designers as power brokers, designers as framework makers, designers as technology enablers, designers as community builders. The latter can be considered as falling in the domain of possibility and inquiry: i.e. the suggestion that designers act as cultural catalysts, as humanisers and as friendly challengers. I would not necessarily say that you are missing roles, but I would argue that there are many variations on how these roles get enacted and combined. For example, throughout our Designmatters projects, I have seen these roles be fluid and shift organically depending on the organisational context, the particular stage of a project's development, and the opportunities/challenges at hand. Our recent project in Chile with Coaniquem, an award-winning medical centre for children recovering from severe burn injuries is a case in point.[2] The brief challenged our students to infuse the environmental footprint of the medical campus with innovative spaces for healing and play. At different stages of this project's collaboration (field research, co-creation with the staff and patients, prototyping, field testing, etc.) all of the seven roles were prominently manifest and paramount to creating impactful outcomes and a very transformative learning experience for all.

With regards to the first part of your question, about which roles resonate the most, I have seen two of these roles, designers 'as framework makers' and 'designers as cultural catalysts' recur with particular force in design teams I studied at IDEO.org; Frog Design, Mindlab and the former Helsinki Design Lab.[3] One of the interesting insights from that research is how often these were roles that were not necessarily codified and expected from the onset of a project. One of my observations was that their emergent nature is one of the reasons that there is an overall sense of an 'uncharted territory' in which designers are operating as they traverse complex organisational boundaries. This is a sentiment that also surfaces with clarity from the perspectives shared in the recent LEAP Dialogues publication, which surveys US-based practices and career pathways in design for social innovation (Amatullo et al, 2016).[4]

Your PhD research is focused on understanding the value of design in social innovation and using the 'Design Attitude' construct to explore this question. Can you explain what 'Design Attitude' is, why you've used it and how it has been used?

My central motivation upon enrolling in the PhD program at the Weatherhead School of Management, Case Western Reserve University, was to address gaps in our understanding about the value of design as a change agent in social innovation. The design attitude construct was for me a way into this journey of inquiry.[5] I was quite fortunate since studying at Case meant having the opportunity to be exposed directly to the thinking and scholarship of the two individuals who coined the term: Richard Boland and Fred Collopy. They defined design attitude as 'expectations and orientations one brings to a design project' highlighting designers' capabilities as a distinct set of heuristics that allow designers to remain in a 'liquid state of becoming' (Boland and Collopy, 2004, Boland et al., 2008). This is a problem-solving approach that deviates from more linear aptitudes for decision-making of managers; it also draws attention to the agility of design under circumstances of uncertainty and complexity.[6] Boland and Collopy's work is part of an important stream in the design and management literatures that call attention to the intrinsic and strategic role of design in organisations. My research attempts to extend their original and holistic conceptualisation of design into the social innovation field.

Because my dissertation relied on the interpretation and analyses of three independent field studies organised in a multiphase mixed methods exploratory design sequence, design attitude functioned as an effective bridging leitmotif for its narrative arc. The design attitude construct was also core to my quantitative work. Here, the research of one of the authors of this publication, Kamil Michlewski, was absolutely foundational.[7] Michlewski had used rigorous, grounded theory methods and many years of immersion in organisations that have high fluency of individuals with design attitude (such as IDEO and Phillips Design). He was able to identify and articulate five key dimensions of design attitude (Michlewski, 2008) that I in turn further operationalised in my work. With some nuanced variations, the field survey of managers and designers that I conducted tested Michlewski's five dimensions of design attitude (my operationalised constructs are 'ambiguity tolerance', 'creativity', 'engagement with aesthetics', 'empathy', and 'connecting multiple perspectives') and established the content and predictive validity of the construct. The study not only showed the positive, significant relationships between the multidimensional construct of design attitude and social innovation project outcomes, but also between two other key variables that we typically contend with in many social innovation projects: the concept of team learning and that of process satisfaction.

What have been your key findings from the research and how does it help us understand the value design brings to the social innovation space?
I was able to triangulate the findings of the quant study with ethnographic research at UNICEF. This allowed me to probe further some of these insights in an organisational context as part of the last empirical study of my

dissertation.[8] In basic terms, the key findings of this research amount to a set of integrated insights and foundational metrics. These findings shed light about the value of design as well as some of its limitations when it comes to its adoption and integration in organisations.

Specifically the quant study validates design attitude as a newly operationalised construct. This is important because we now have some valuable data that allows us to articulate with new precision key design abilities and approaches to innovating that are significant in accounting for success. Common design techniques—prototyping and visualisation and manners of practice-user participation in the design process—are also constructs that are captured in the study and analysed for observed impact. All too often our qualitative-based studies in design struggle with the question of measurement when it comes to the impact of designers' modes of engagement in the field of social innovation. This research presents strong evidence about the positive role of design attitude in social innovation project outcomes, demonstrating with new empirical data the value designers brings to the social innovation space.

If we agree and accept that a design culture/attitude is key to helping an organisation's overall transformation, how do we evidence it and make a compelling argument for it? Is your 'Return on Design' (ROD) conceptual framework an attempt to address this?
The 'Return on Design' (ROD) framework that I offer in the dissertation posits design attitude as a generative force. It recognises how design can be associated with notions of stewardship and collective agency, which are known to transform organisations. However, how much design attitude proves transformative depends on the particular organisational culture at hand, and at what level of the organisation (micro, meso and macro) it is present. In addition, how the abilities and dispositions of a design attitude are experienced and perceived is likely to vary widely depending on the worldviews of organisational actors.

What are the enablers and barriers to promote and integrate a design culture as part of an organisation's overall innovation agenda?
This is a difficult question to answer in a conclusive manner. Based on the insights of my research, I would offer that many key enablers for promoting a design culture that will ultimately result in transformation, align with leadership buy-in and the extent to which the norms and practices of a design culture get integrated into core organisational principles and values. In addition, the wins of design in an organisation also seem to strongly correlate with the level of entrepreneurialism of its human capital across divisions, as well as with the organisation's overall orientation toward collaboration and change. On the opposite end of the spectrum, among the key barriers

I identified are notions of accountability and legitimacy that are connected to larger institutional logics. These can often lead to risk-aversion and a lack of creative confidence that have negative implications about how design is perceived and understood in the organisation at large.

Many of our readers will not have experienced the type of 'design' we are espousing here. How do you talk about the role, value and impact of design to a non-expert? Do you think the definition and description of the seven roles will help?

I always remain inspired by Richard Buchanan's seminal articulation of design's four orders since they offer a very tangible way to talk about design's impact in the world across communications, products, environments and organisational systems (Buchanan, 2001).[9] I also adhere to his observation that design is a humanistic discipline without a subject matter, and therein lays it power and challenge. There is no question that the transformative role of design in organisations can be abstract for the non-design reader. The seven roles that this publication articulates are helpful because they illustrate in concrete ways how designers practice and what they do. The scenarios and examples offered are ones that everyone can relate to, whether they are familiar with design as a knowledge discipline, or not.

And finally—given the subject of our book, what is the one key point that you will want to take away from the book and why?

I anticipate that this book will continue adding insights to a growing body of literature and case studies that reveal, with powerful empirical evidence, the unique value professional designers impart to processes of social innovation and transformative organisational learning. One key take-away for me would be the potential the book has to counter some of the technocratic emphasis we often encounter in many streams of the current innovation discourse, and instead reinforce the humanistic strength design brings to this field of inquiry and practice!

Endnotes

1. Designmatters was established in 2001 as a college-wide program of highly curated educational projects and publications during the presidency of Richard Koshalek as part of an institutional vision to bring an infusion of innovation and global engagement into the organisational fabric of ArtCenter. During the program's first two foundational years Designmatters co-founder Erica Clark and I collaborated with a task force of staff, faculty and alumni to develop the program's mission, core principles and outreach strategy. Key inspiration for the program came from grounding research conducted by Clark of social impact design initiatives in a select number of European Institutions, including a program led by Liz Davis at ENSCI, Les Ateliers, Paris. In 2008, Designmatters transitioned from a college-wide initiative

to a full-fledged educational department with integrated curricular responsibilities across the college. For further reference about the early history of Designmatters, see Mariana Amatullo and Mark Breitenberg, 'Designmatters at Art Center College of Design: Design Advocacy and Global Engagement,' Cumulus Working Papers, Nantes, 2006, and Mariana Amatullo, Erica Clark and Richard Koshalek, 'Design Education for International Engagement', in proceedings of the Inaugural International Forum on World Universities: The Role And The Future coined the term of The University In A Changing World, Davos, Switzerland, January 31-February 2, 2008.

2. For further information about the Designmatters collaboration with Coaniquem and case study see: www.designmattersatartcenter.org/proj/safe-ninos/

3. My research question in that study interrogated the influential factors that define the role of the designer in the social innovation context. The cases examined include projects addressing human needs linked to large-scale social, cultural and economic challenges. For more background see chapter 1 in my dissertation, 'Design for Social Change: Consequential Shifts in the Designer's Role', in Mariana Amatullo, Design Attitude and Social Innovation: Empirical Studies on the Return of Design, Weatherhead School of Management, Case Western Reserve University, Ohio, 2015.

4. Mariana Amatullo, Bryan Boyer, Liz Danzico and Andrew Shea, ed. LEAP Dialogues: Career Pathways in Design for Social Innovation, Designmatters and Distributed Art Publishers, 2016.

5. Mariana Amatullo, Design Attitude and Social Innovation: Empirical Studies on the Return of Design, Weatherhead School of Management, Case Western Reserve University, Ohio, 2015 (dissertation chair Richard Buchanan; dissertation committee: Richard Boland Jr., Kalle Lyytinen and John Paul Stephens).

6. Boland and Collopy argued that managers should not only act as decision makers, but as designers, and called on managers to learn from designers' open orientation to projects, from their treatment of situations as opportunities for invention, from their questioning of basic assumptions, and from their resolve 'to leave the world a better place than we found it' (Boland & Collopy, Managing as Designing, 2004: 9).

7. I am particularly indebted to my dissertation chair Richard Buchanan for introducing me to Kamil Michlewski's article, 'Uncovering design attitude: Inside the culture of designers'. Organisation Studies, 29(3): 373-392, 2008. This research and subsequent dialogue with Michlewski informed the design attitude scale that I developed in my quant study with the mentorship and close guidance of one of my advisors Dr Kalle Lyytinen.

8. For a further synthesis and extension of the findings from the UNICEF study of the dissertation, see Mariana Amatullo, 'Insights from Integrating a Design Attitude Approach to the Innovation Ecosystem of International Development', in Proceedings of the 20th Design Management Conference, Design at the Point of Inflection, Design Research Meets Design Practice, Design Management Institute, Boston, 2016.

9. Richard Buchanan. Design research and the new learning. Design Issues, 17(4), 3-23, 2001.

Public sector innovation through design

Brenton Caffin

Brenton Caffin has been the Director of Innovation Skills at Nesta since 2013. Prior to this, he was the CEO of The Australian Centre for Social Innovation (TACSI) which he founded in 2009. He has over fifteen years experience working on public sector reform, half of it spent as a public servant and half of it spent externally as a consultant. He has always been interested in trying to discover and provide the mechanism and methodologies for creating change in public sector organisations. As a classically trained economist who has spent the last six years engaging with design and designers, Brenton's main role has been to translate the capability set of design in the context of public administration.

How has the landscape of social innovation changed?
The development of social innovation has been very different in Australia, Africa, Europe and America and as a result we have seen development in a range of different directions. Over the years, I have seen formal institutions take greater interest in a design-led approach and they are more accepting of different ways of doing things. For example TACSI was set up as an experimental lab outside of government five years ago. Now, there are many innovation labs springing up inside of government because there is increased recognition in the importance of having these core skills and competency in government. These labs are seen as more credible now and we are seeing the relationship between formal institutions and start-ups changing. One of the biggest changes I have seen so far, is the way UK is developing a much stronger ecosystem to support these types of activities. The ecosystem is creating a platform to help identify and accelerate the development of ideas through various funding mechanisms (through networks) and providing structural support like creating working spaces and social impact bonds. These support structures were very nascent five years ago and now they are at point where the ecosystem is becoming well established and has become a point of reference and replication around the world.

What differences have you observed between the Australian and the UK/ Europe context?

The UK social innovation landscape is much more mature in the UK than it is compared to Australia. It might be down to scale and size of the market, but there has been fundamentally larger and continued investment into the social innovation ecosystem in the UK. Whereas in Australia, development and investments has slowed down and as a result organisations like TACSI are still mostly seen as the exceptions rather than the norm.

Perversely, the investment climate in Australia has got tougher despite the fact that the budget situation in Australia is not nearly as bad as it is in the UK. As the public spending cuts worsen in the UK, innovation has become part of the solution of how you 'square the circle' by maintaining or improving services with fundamentally lower resources. There hasn't been the same existential crisis in Australia and the public sector has not had the same pressure to deliver better services at a lower cost. As a result, innovation is seen more as a luxury than a necessity. Although the Australian local governments recognise the need for change, they have not personally 'felt' the pain of cuts. They haven't got the coals under their feet quite as firmly as they do here in the UK.

Additionally, a lot of our activities at TACSI were mostly about educating our partners about the different ways of creating new programmes and developing solutions, as well as educating them about the different ways of procuring support to redesign services. These spaces did not exist previously and we had to do a lot of the heavy lifting at TACSI. That has been one of the key differences between the UK and the Australian context.

What are the new capabilities leading the transformation of the social innovation sector?

I want to highlight two areas. The first is the advocacy-related skillsets required to bring impact to this space. Although different organisations have slightly different approaches and emphasis, broadly there is a recognisable range of skillsets required for this type of work. The capabilities that I am referring to are not just limited to doing the work, but also around the advocacy activity to transform this sector. It's not just about finding funding for small projects, but using these projects as a means of persuading people to work in a different way.

Secondly, I am also seeing a growing ability within the social innovation sector to engage with large organisations and to help them to adopt some of these methods and approaches themselves. It's about helping organisations to begin embedding these skillsets into their own practice rather than funding this activity at arm's length.

What is the role of design in this context?
A cultural catalyst

The role of design has been to be a catalyst to a different way of seeing, thinking and acting. Design has the potential to give people an alternative set of tools and an alternative way of seeing and shaping the processes that they undertake. It helps them to refine and evolve the services that they deliver on behalf of citizens. I always like to challenge the top-down method of policymaking and design of services. Design methods enable people to do work from the bottom-up and really understand whether or not their new proposition is actually attractive to citizens. If we prototype early we can accurately test if this new service or policy works before millions of pounds are invested. It's a useful and cost-effective way to manage risk.

A humanising aspect

We spend a lot of our time working with early adopters. My job is not to persuade the ninety-nine percentile that they need to stop working the way they are. We are working with a growing number of people and organisations that have the appetite and willingness to try something different. They recognise that doing things in unexpected ways can be a way forward. However what they don't have are the tools to do this. This is where design has really opened up a completely different way to view and think about public sector innovation. Christian Bason (from MindLab and now the Danish Design Centre) basically talks about this in terms of the analytical versus the emotional. The public sector has been traditionally dominated and driven by disciplines that are predisposed to using the left hemisphere of the brain and, I think, design brings a little bit of the right brain into the mix. I don't subscribe to design being the saviour of everything but I do think it can be a really healthy complement and offers an original approach to help people question their assumptions. It helps validate ideas by testing them through prototyping so we better understand how it will be used.

What in your opinion is the most valued aspect of design in this space?
Offering a framework

The most valuable aspect of design is, for me, showing public sector workers how using simple techniques early on, like shadowing users and observing how they respond to the service through touch points, and artefacts can be used as props to help them improve the services they are delivering. It is about giving people the confidence to build ideas through prototyping and seeing how it is received, rather than trying to rationalise it first on paper. There is a strong orthodoxy in needing to be cerebral and logical in existing public administration models. It is very much about creating this idealised 'world' and trying to make the real world conform to it. Design in contrast valorises the idea of trying things out. It's ok to leave our desks and spend time walking up and down the high street, observing people. It's ok to knock on doors to have conversations with people as a source of inspiration. So going back to my earlier point, design has really given people the tools and process to help them develop their services differently.

Design also places strong emphasis on prototyping and the expectation that things will not be right the first time round. It is about testing early and refining it many times. This is the direct opposite to what you expect from a traditional policymaking process. It is really important to embed these innovation loops and expectation of rapid innovation into the current process.

What have been the challenges in bringing a more design/creative-led approach to this space?
Managing change
The most difficult challenge has been in trying to embed this new way of working. Some people may understand it at a conceptual level but they find it difficult to embed it in their practice. There is a concept called the 'zone of proximal development', where if you are trying to encourage people to adopt a new skill or behaviour, learning mainly occurs when you push the learner out of their comfort zone. It's similar with muscle development, you're trying to push people beyond what they are used to but not so much that they completely disengage. When you are working with people who haven't worked with these methods before you have to work really hard at creating space for the development to happen otherwise you run the risk of falling at the first hurdle. The skill of diplomacy and the tact to guide people through this change process is probably the most difficult skill to find and nurture.

Finding the right people with the right skillsets, nurturing them and creating a career path for them
It's been a real challenge finding people with a balance of design skills with other softer skills (such as diplomacy and advocacy skills). It is incredibly difficult to find, even amongst senior designers, someone who is capable of going in and talking with CEOs. And because of this, it is important to create a flow of talent.

A visible career ladder in innovation is still missing from government. And so we often end up with people who are attracted to the role because the job title has the word 'innovation' in it and that is quite worrying. Often people who apply because of that are the people who you would least want to put in that role. In contrast, you also get people doing innovative stuff but covertly in order to get around the system. It is very difficult to raise their profile because it works counter to the way they have achieved change in the past. It is really important for these organisations as they mature that they do find ways to integrate these skills and methods into the way they do business. Simply procuring at arms length will not work.

Understanding the process

There is an inherent tension in the design process because design always challenges the brief. Designers have to be tactful in challenging assumptions. Public servants who have commissioned the work probably do not take kindly to being told that the problem they have identified is wrong. And so, there is a real challenge in making public servants more educated, familiar and comfortable with these sorts of processes.

Understanding the funding landscape and culture

Its also really important to have sufficient literacy and knowledge about who the funders are, whether they are commissioners of service in the public sector, grant makers or if they are philanthropic. On the demand side, it is about continuing to provide more evidence and proof in the value of working this way, and making it easier for organisations to adopt or procure these kinds of services in a more effective way. In comparison, on the supply side, it is about getting the talent, which I have highlighted earlier.

How to engage with the uncontractables?

One of the key challenges has been trying to build internal design capability and to entice designers into government. The recruitment processes of government agencies and multi-lateral institutions that I have worked with in the past often do not have appropriate ways to attract and engage with these new talents. Giulio Quaggiotto coined the term 'uncontractables' to describe the challenge of recruiting these talents into the UN system. They are 'uncontractables' because it is often really difficult to get designers interested in joining large, bureaucratic organisations. Its important to recognise that some of these institutions may not be the most attractive places to work for and its important that we think of different strategies to engage with designers in more informal ways, for examples through collaborations or residencies rather than through contracts or permanent roles.

Building credibility

It is also really important that we use our work to help us build credibility in the approach and methods. For example, we funded a project internally at TACSI because we did not want to spend twenty-six months trying to raise funds for a project that we didn't know would work. So we explored questions relating to families without knowing what the outcome might be, and as a result we arrived at a process that worked and used it to make the case for the next project about ageing and carers. We had to find ways to build credibility and demonstrate a proven process.

What are the optimal conditions for design to be effective in this context?
Political leadership, resources and right mix of people
My experience in TACSI has shown that it's important to have a clear authority from the political leadership so no one would question our right to operate. It was also about having the resources to deliver a very high quality outcome that can be used to demonstrate its effectiveness. And the final element is to have the right mix of talent, so for example, having someone from design and social sciences working with public servants seconded from government. It needs to be a team that can straddle multiple sectors, and to have people who can play that diplomatic role in keeping an open space to allow a different approach to work. It's the combination of these three things: authority, access to resources and right mix of people supported by a political platform that are key for design to work in this space.

Design in government
Christian Bason

Christian Leads the Danish Design Centre (DDC), which works to strengthen the value of design for business and society. Prior to joining DDC, Christian headed MindLab, a cross-governmental innovation lab, and the public organisation practice of Rambøll Management, a consultancy. Christian is also a university lecturer, and has presented to and advised governments around the world. He is a regular columnist and the author of five books on leadership, innovation and design, including 'Design for Policy' and 'Leading Public Sector Innovation'.

How did you get involved in design-led innovation in government?
I graduated with a Masters in Political Science and I was hired as a consultant in Rambøll Management, one of the largest Danish owned management consultancies. After a few years, I was asked to lead the public policy evaluation and analysis team.

During this time, while working on a strategic project commissioned internally, I became intrigued by the notion of innovation in public sector. We interviewed over twenty-five senior top-level public managers across Denmark to understand what public sector innovation meant to them. We held a conference and published a report based on what we found out. As a result, I was approached by the Danish's Ministry of Business to join MindLab as a director. This was where I was introduced to design.

What is MindLab and how has it developed?
Background
MindLab was originally set up in 2002 by the Ministry of Business Affairs as an internal incubator for creativity and innovation. It had a core team of five staff and was considered to be one of the world's first public sector innovation labs. Its original aim was to improve the efficiency of the policy development cycle by shortening process times especially in cross-organisational policy processes. Although the experience with MindLab as a facilitation unit was very good, by mid-2006 the current operating model had run its course, with creative thinking being more widely accepted and the shift to project based portfolio work. There was a need to review its services and aim. There were also suggestions and on-going talks to locate MindLab across other ministries, particularly the Ministry of Employment and Ministry of Taxation. So when I joined MindLab in 2007, the mandate was to develop a new strategy

for MindLab and to integrate these three different ministries. I arrived at an opportune time. The existing MindLab model had run its course and many of its staff had already left. So I was given a year to redesign the political space and to get a revised MindLab up and running with a new team, strategy and model in place.

MindLab 2.0

The revised MindLab became a cross-governmental innovation lab and its revised aim was to work actively to promote innovation in the Danish public institutions through inter-governmental collaboration on user centred innovation in policy and service. The MindLab core team had backgrounds in political science, interaction design and anthropology, and only two members had longer practical experience as civil servants. We have proven since it was revamped in 2007 that it is a sustainable model. We have had the Ministry of Education joining and other pilot programmes with the Municipality of Odense and the Ministry of Interior. MindLab is also generating income through keynotes talks as well as providing consultancy service to foreign governments and cabinet offices.

A learning organisation

I have always intended for MindLab to be a reflective and learning organisation. The original team also included PhD students recruited for each of the three different ministries. This was an important part of the strategy as it created an environment where we could reflect and learn. From the beginning, I was conscious that we needed to document and share what we learnt by building up case studies. We spent a lot of time sharing our stance (through talks, reports and books), which helped us reflect and refine our understanding of what we were doing. I believe this is one of the reasons why the MindLab story is well known in the sector because we reflected, wrote and shared a lot of what we did.

Establishing an innovation lab in 2007 was really new territory for the Danish government. We were also considered to be one of the first governmental innovation labs. Since then, innovation labs have sprung up everywhere. For example, over three hundred people from around the world came together to discuss innovation labs in government at Nesta's LabWorks 2015 in London. I wouldn't go as far as to claim that this visible growth is solely down to MindLab, but we have definitely been a key part of that journey.

Why has there been a growing interest for using design in the public sector?
There are a number of reasons why design is increasingly being talked about. Firstly, it has to do with context. For example, the context of the global

financial crisis has led to cuts in public funding. In Denmark, an ageing society and a lack of skilled workforce have placed additional pressure on public services.

Design is a profession that is concerned with the creation of novel and new ideas. It's not surprising that it is increasingly being sought to help find solutions to complex social problems.

There is also an alignment of values between designers and public managers. Both are generally interested in changing people's life for the better. For designers, its traditional through the designs they created. For politicians and public managers, it is through policy.

The word 'innovation' generally invokes the idea of open experimentation. However it does not necessarily lead to any real change. Its becoming extremely evident as I write my PhD, that design is not only good at offering a set of methodologies, approaches and tools but also a set of values that can drive actual change.

How is design used to catalyse change?
Through the fifteen case studies used in my PhD analysis, I have identified three key drivers of change.

One is user insight, or the empathy in ethnography, the ability to understand user experience. The second is systematic ideation and creating ideas as a driver. The third is visioning, that is the ability to imagine a future state as a key catalyst of change.

For me, the most powerful drive or trigger of change processes has been insight. It has been the empathic quality that design brings and the ability to orchestrate and curate emphatic materials (such as videos and audio) into clear directions for innovation and directions for change. This quality has been key for MindLab.

Design is also really valuable to help us reframe the world and our assumptions about it. Sometimes it's challenging assumptions and in some cases it might be confirming hunches, but being able to make it concrete through visuals or films can be very powerful. It makes it hard to avoid. I have been approached by managers who say things like: 'Ouch!', 'That was painful' and 'That hurt'. But I've also heard managers say it has been great.

The nature of the design process empowers implementation. The focus on testing out new ideas quickly and iteratively gives confidence to the project team. This is a key difference between design and other management consultancy. Management consultants stop at the research and analysis stage, while a design process offers support and tools to help develop the idea through to implementation.

What are the key challenges in introducing and sustaining design-led innovation?

Building legitimacy

One of the key challenges has been to build legitimacy in a context that is characterised by bureaucracy, politics and a lot of complexity. It was important for MindLab to have a mindset that is not afraid of challenging existing assumptions but to do so by understanding and engaging with the system that you are part of.

Working with the context of the organisation

I also wanted MindLab to challenge the organisation, to be its 'conscience'. After the first two or three years of growth, I became increasingly concerned that we were too occupied with ourselves and might be in danger of becoming out of touch with what was happening on the ground. I distinctly remember saying at a strategy seminar in 2010 that we had to challenge ourselves to reconnect back with the organisation.

Ensuring relevance

A way of ensuring that we are delivering public value is to constantly ask ourselves the 'so what' question. It's a question that probably isn't asked often enough and is really crucial in helping see beyond the flashy workshops, cool processes and massive visualisations. It's really important that we maintain an appreciation and empathy with the organisational context that we are trying to catalyse and have an impact on. It's really critical and, in my experience, designers sometimes lack the tools or even empathy to understand the systems in organisations. So being aware of this and constantly asking ourselves the 'so what' question is important.

Evidence of impact

We also have to be careful in the claims we make about design and its potential impact. At the Danish Design Centre (where I am now based), our aim is to help decision makers identify the most appropriate context and problems that design can help with. It would undermine our legitimacy if we were uncritical of design. It is really important for us to identify what are the 'fingerprints' of design contributions and what are the artefacts that will facilitate new kinds of dialogue.

Engagement

Earlier, we spoke about why design fits well with designing public services and policy. However I do think there is a paradox. Designers have a very human instinct and love to think of themselves as dealing with social problems. But they don't see themselves as being part of the government that has the responsibility of improving its citizens' lives. It's a struggle for many designers to keep their freedom and operate separately in their own studio and still be part of the conversation.

What are the optimal conditions for design to be effective in this context and how do you encourage these conditions to happen?

Drivers of change

For a start, it is really important to have pre-existing drivers for change. This could be brought on by an austerity drive to reduce costs and achieve efficiencies or finding new ways to meet increasing demands. Sometimes it is to respond to citizens' complaints or finding new ways to achieve the kind of impact they want.

Bridging disciplinary divides

It is really important to be able to speak the language of the sector that you are working in. As someone with a political science background, I have a nuanced appreciation of what a government is, what it's trying to achieve and what the challenges are. I am able to empathise with other political scientists and public managers through my previous experience. Being able to draw on my background and experience has helped me build the bridge between design and public policy.

Legitimacy

In my ten years of consulting, I have used many different types of qualitative and quantitative research. While in MindLab, we mainly used design ethnography and other design research methods. Having experienced both 'traditional' and 'design' research methods, I am able to objectively identify and evaluate the differences. I have been there and done it, and this gives me a level of legitimacy. There is legitimacy to my promotion of design since I did not come from the profession. Being an outsider (who has experienced both worlds) made it more legitimate for me to highlight to my fellow public managers why design is interesting.

Navigating discomfort

Design (as I said before) is really useful to help us challenge assumptions. But it is hard and can be painful and stressful for staff. So I have a concept that I refer to in my PhD, which is the concept of 'navigating discomfort'. It is really important that public managers and commissioners of design take on the role as 'nurses' and support staff in navigating through the difficult change process.

Being open, opportunistic and challenging assumptions

The role of the manager can't be underestimated and its really important they have a set of attributes that enables the project to flourish. A key attribute observed in successful public managers is a propensity to challenge their own assumptions and to ask disruptive questions of themselves and to their own organisation. They are often comfortable with change and are open and curious to find new ways to create meaningful change. It helps that many of them have had careers that are in very different fields and go beyond the conflicts that they are currently experiencing. Some of them have unusual educational backgrounds and many of them, I believe, have a value set that is focused on change. They are also opportunistic and not afraid to try new approaches. If they think design can help them achieve change, then they will use it. I can see how they become quite enthusiastic and energised by design because design supports them to create meaningful change. Recognising and seeing evidence of these attributes is really important, to the point where if we think the manager is not sufficiently open enough and prepared to join the process, we would actually take the radical decision to stop the project.

Senior management support

Having the support of senior management and the ability to level the playing field is also really important. It is also to enable staff to have the space and time to engage. For example, in a project with the Danish National Hospital, we conducted an ethnographic study with twenty heart clinic patients to understand what their experiences were. Some of the patient stories were very challenging to hear for the staff but we managed to achieve significant changes, partly because we had the support of the head doctor and nurse. The two senior managers were also very hands-on and participated in all the workshops. It was important to have the workshops in a neutral space outside of the hospital premises, since it meant that everyone participated on an equal footing. Everyone came in civilian clothes and the confines of their professional identity were left behind. In this particular case, we managed to redesign a physical space, radically changing the way patients and staff interact, changing the entire workflow and work processes to make them more patient centric. Part of the project's success was attributed to having the permission, space and time to challenge and change, authorised and legitimised by senior management.

Alignment of values

For a design project to be successful, ultimately the findings or the insights have to be in line with the mission of the organisation. If the design project ends up being directly at odds with the overall political direction and intent of the incumbent government, then it will not progress. To get around this problem, many projects tend to operate in 'stealth mode' to enable further development before they are scrutinised through political agendas. Ultimately, good design work exposes whether the organisation is doing well or not, which makes design really valuable but also challenging for governments.

Digital transformation
Beatriz Lara Bartolomé

Beatriz is an expert in innovation and digital transformation, a pioneer in collaborative work at large corporations and a visionary amongst her peers. She is currently the CEO and an investor in Imersivo, a start-up focused on bridging the online and physical retail experience. Beatriz has many years of experience working in a highly innovative and technological telecommunications sector where she served various roles in management positions at ITT-Nokia, AT&T Network System, Ericsson and Alcatel. She brought her expertise to BBVA–Spain's largest bank where she served initially as the Chief Innovation Officer and then as the Global Director of Corporate Transformation until July 2015.

How did you become an innovation and digital transformation expert?
I was born innovative. When I was a child, I wanted to be an astronaut–you can imagine it wasn't so common for a girl 45 years ago. I am always learning and reinventing myself from when I first started out as a researcher, then as a corporate executive and now as an entrepreneur.

I used to say that when someone isn't able to find the right person to launch a new initiative or programme in the company, they would call me. So in many ways I have always worked in jobs that were highly challenging. Jobs that you could say nobody wanted. How did I get there?

My educational background was in physics and my first job was in a factory that was bought over by Nokia in the mid-eighties. That was my first contact with mobile telephones and the start of a long career in the telecommunications industry. I then worked at AT&T to re-define the information systems for a just-in-time manufacturing approach. Then there was the introduction of the GSM digital standard. At that time, there were different standards adopted in the US and in Europe. AT&T needed to adapt their systems to offer GSM signals for the European market. Again that job fell to me. I knew when working at AT&T that the future of telecommunications was in mobile and so made the decision to move from the largest telephony company in the US to a traditional Swedish manufacturer, Ericsson, because they were pioneers in GSM technology. And from there, I moved to Alcatel where I was in charge of Mobile Operators for Latin America, and their Intelligent Network Competence Center for the design and development of Mobile Applications (5 years before the App Store creation).

In 2006, I made the decision to leave the telecom industry to work at BBVA bank because they offered me the possibility to be a pioneer (once again) helping them research and implement emerging technologies that will impact the banking industry. I created BBVA Innovation Center and led the migration to the Google Cloud Platform.

Why are current organisations finding it difficult to innovate?

One of the reasons is because they have people in the executive team that do not fully understand the complex technology behind their business. My observation is that CEOs and EVPs of large companies in various industries still consider technology as a business enabler instead of a business driver for growth and differentiation.

In the second half of the 20th century, companies have naturally adopted the Information and Communications technologies for their back end systems and processes. However, in this century, the SMAC Technologies (Social, Mobile, Analytics and Cloud) are changing the relationship model with customers, and are becoming part of a company's front-end systems.

One of the most important things I learnt during my time at Ericsson is that a leading company should have technical people in the top positions of the company. Innovation requires companies to be at the technological edge all the time and you cannot do that if the top management are not advised by people who understand the technology, and its impact on people's lives.

What is digital transformation?

SMAC technologies are great catalysts of change and facilitate exponential growth. They have changed our lives and the world around us. We have become prosumers in the new sharing economy. That which did not exist some years ago is now essential to us. The Internet is no longer a 'new channel', it is the world we live in. Companies have no other option but to move to this digital world in order to remain relevant for their customers in the coming years.

The digital transformation of an organisation is the result of a gradual adaptation to the new rules, possibilities and opportunities of the digital world. This digital challenge implies the adaptation of the relationship model with customers and employees, and the resulting business processes and procedures. A mindset change, new competencies, and new ways of working are also needed to achieve the goal. Real digital transformation occurs when companies are able to seamlessly move between the two worlds in the relationship with their customers and in their internal operations.

Ninety per cent of companies are not digital native. As a result they transfer what they know from the physical world to the digital world. Digital transformation is not easy in the sense that the existing managers and employees are used to dealing with the operations of the physical world. People used to this way of business are not always able to translate what they know into the digital world. As a result, you get two internal companies running alongside each other that do not interface well at all.

Are there any good examples?

Yes. Take for example Ikea. One of the Ikea apps allows you to use your phone camera to place a piece of furniture from the Ikea catalogue into a room and location that you had in mind. It's a great experience. Ikea is really good at controlling this transition between the physical and the virtual. Ikea's front-end systems are also well synchronised with their back-end systems to deliver the selected product with all its assembly pieces.

How do you digitally transform an organisation?

'Building the new normal' stage-by-stage is my approach to digitally transforming an organisation. In this way, we can define and build a digital journey at the corporate and individual level.

Becoming digital

The first stage of digital transformation is when a company takes the decision to become a digital company, or realises that it needs to. It's not an easy decision to make but once it is made the changes need to start from the inside. Digital transformation is part of a broader innovation portfolio. In my opinion innovation tries to really understand what is going on outside in order to understand how you should change from the inside. It won't work if your employees are not on board. The change should start from people, the employees. They should adopt the change if you want to impact the business. This means that the next step should be about putting in place the infrastructure to enable employees to change. For example, changing the company's internal information systems or workplace technology. The new world is not only about a new relationship model with our customers, but it's also about a new relationship model with our employees. Employees of the company need a digital identity they can identify with and having the infrastructure and additional support will help them be comfortable with technology.

Behaving digital

Once you have defined what it means to have digital employees, you need to enhance their experience and help them take responsibility and ownership of their work. Imagine there is a red line. On one side of the line is what a company does for their employee–make decisions, provide the right infrastructure, facilitate learning. On the other side is what employees are able to do. Learning is really important in the transformation process. Learning is an adoption process. In today's world, it is not just a question of being more effective or productive; it's a question of being more flexible and adaptable.

These are key survival strategies for large companies. So the next stage in a digital transformation process is Collaboration. Once people have changed individually, they can work differently with others. This is not always easy to achieve in large corporations. Managers supervise tasks instead of providing vision and guidance. Colleagues are often seen as competitors, vying for the same resources. Hence it's important that we learn to collaborate internally and create a networked enterprise. If we are able to achieve this stage, then it means that we will achieve business impact by being customer-focused. People are really important and ultimately people not robots deliver services.

What are the challenges of digital transformation?

The challenge of going digital at scale is its complexity. To mobilise an organisation towards the wanted position, it is important to consider all its perspectives. Roberto Fernández, a Professor of Organisation Studies at the MIT Sloan School of Management, describes three perspectives: strategy, politics and culture. The digital journey is on a road of two lanes: towards customer experience, and towards employee experience.

Customer experience: Seamless transitions between the physical and digital

Companies often make the mistake of thinking they are digitally enabled while they are really running two companies in parallel. They often transfer their current competencies into the new digital arena. They start by creating a new website, offering transactions on their website, focusing on usability aspects and creating social media platforms. It certainly helps enrich their brands but it is not really digital transformation. They may have a good presence in the digital world but they often end up with two different experiences for their customers. A successful digital transformation should allow the company to control the transition between both worlds, from the physical to the digital and from the digital to the physical. The point is that you move seamlessly from the digital world to the physical one, so once you have captured a customer in either world and to enable them to move between the two successfully, then you are really closing the loop.

Digital transformation does not always mean cultural change

Many companies consider digital transformation as major cultural change and start by putting in place a classical 'change management program', usually led by Human Resources. I don't believe in culture change as a driver of digital transformation; it could be an outcome. I think that the culture of your company can be exactly the same. Imagine your values are respect and perseverance. You can have respect and perseverance in both the physical world and the digital world. The point is that you need to observe these values and behave in a different way depending on the environment. So digital transformation, does not in my opinion, necessarily imply a change of values.

Focus on people: Employee experience

The digital journey implies the adoption of a lot of new competencies. It implies that we must first understand how the digital world works and then start by learning the tools and new methodologies like Design Thinking and Agile. Very few top managers and employees in the large companies are able to configure their own email account and/or print settings by themselves. So the challenge of digital transformation is to improve the digital literacy of the people in the company.

People are the star of the transformation. You cannot provide a service without your own employees. You cannot change all of your employees so buying new companies to acquire critical mass of new competencies is quite common. You need new people with deep knowledge in these new competencies, for example cloud services or gaming, but in parallel you need to train and motivate your existing employees to upgrade their skills.

Why is design important for digital transformation?

Human-centric Design

I like to use an innovation definition from Curtis Carlson and William Wilmot's 2006 book titled Innovation: The Five Disciplines for Creating What Customers Want. Innovation according to Carlson and Wilmot is 'creating and delivering high customer value in the marketplace, also allowing sustainable value creation for the company'. This means focusing on the customers and their needs. If you really want to be customer-focused, you really need to orientate your company using a human-centric design approach, and you need designers who understand how to be human-centric on your executive team.

A great majority of large companies, associations and institutions are generally managed by business administrators. They focus on the outcome of the operations (figures and financial ratios) but they tend to underestimate or take for granted that these outcomes are generated through day-to-day operations of the business by people for people. And that means focusing on understanding people, their lives and aspirations, and having a vision and an understanding of the technologies as well.

There are very few organisations that can really achieve a global mindset. To think global in my opinion is not just that 'I have a product and I can sell it globally'. Thinking global is to understand that not everybody has the same standards of living and that not everybody has the same values. However we do basically have the same needs with different approaches. So if you want to really serve the customer first you need to be human-centric in your approach and then you need to consider their culture. You will also need anthropologists (as one of the design profiles) in the team to understand the differences between countries and cultures if the company aims to be global.

Bridging different worlds

One of the ways to help experts in technology and experts in business connect is through the language of design. Why design? Designers provide the bridge because designers focus on the customer and not what the technology is capable of doing or how it fits in the existing business model. They say 'no, no, no, don't look at each other, but look at what the customer needs'. I would recommend that large companies hire designers and to have at least one designer at the top level of the company. It's a way to connect the operational world of today with the operational research of the future. It's a way to connect technology with business. It's a way to connect a physical world with the digital world.

What are the key challenges in using design?
Unleashing creativity within the organisation (paraphrasing Tom and David Kelley)

Design is not just a process. It is a critical discipline to create and/or maintain a company's competitive business advantage in a changing world. Creativity is the soul of design. Even though everyone is a potential designer (creativity is part of human nature) not all top managers are professional designers. And so a key challenge for companies using design is not just to follow a prescribed design thinking process but also to take into account the one per cent of creativity. It's not about following a design process rigidly but embracing the creativity and bringing in a lot of people from different disciplines to work together.

Bridging the operational and innovation worlds

When I talk about building a bridge this is what I mean. The operations world is completely different from the innovation world and operates with different rules. Design is the bridge between the operations world of procedures, rules, decision-making and routines. The world of innovation is all about experimentation, curiosity, creativity and connection making. So it's extremely important that both worlds start to talk if we want to build what I call the 'ahaow' moment. The 'ahaow' moment is when you realise the business relevance of what you have in front of you (Aha!) is also memorable (Wow!).

What are the key competencies required for digital transformation?

I have identified a minimum of 12 new competencies that companies need to develop successfully in a new digitally-enabled environment. They are: cloud operations, big data, UX design, enterprise consumerisation, quality engineering, consumer experience, mobile, internet of things, continuous evolution, digital content, cognitive computing and gaming.

Firstly, they need to understand how cloud services work because services and applications are increasingly being hosted remotely due to the cost required to maintain in-house services. Companies also need to develop competencies around how to process data in real time since it enables them to respond to changes in the market quickly. They also need artificial intelligence systems to help them make sense of the data.

Companies will also require user experience design expertise since they need to understand their own company in a different way and to work in an agile manner. This is linked to the concept of consumerisation of the enterprise where systems used internally by employees should just be as usable as systems designed for external customers. And because software development is quicker due to agile methods, it's also important that we approach quality systems in a completely different way. For example, consider the ethical decisions that have to be designed into an autonomous car system. How should a car be programmed to act in the event of an unavoidable accident? Programmers have to evaluate the risk and consequences of machine learning, and the decision rules that they are designing into a programme. Software developers using Agile methodology are not required to escalate or ask permission to create new functions or improve on existing ones. So it's important that they have an ethical approach to coding.

It's a given that companies have to understand mobile. It's extremely important to understand how people use mobile devices and services. Internet of things is not a concept it's a reality. For example, in 2018 all cars manufactured and sold in the EU will have an alarm system that will automatically send an alert to the emergency service immediately after a crash. This is based on research that the first hour of a crash is critical in order to decrease mortality rates in car accidents.

Six years ago touch screens were rarely used. Now it's everywhere. It's important for companies to have the know how to create their own digital content. Companies need to constantly build and launch their own content in different content platforms in order to connect with customers and employees. So digital content generation should be a competency if you are a large company.

This is my summary of key competencies that a company needs to develop in order to be a successful player in this new environment. What ties them all is the importance and role of design. My main point about design is that design is about people. We cannot build a successful company without considering people.

So how do you build these competencies?
Offer a learning journey
We need to understand how to control the transition between the physical and the digital world. Then identify what the critical competencies are for

that particular company or industry. It's important that we don't just design a training programme but a learning programme. It's not about training your employees but encouraging them to learn. You cannot teach if they don't want to learn. And in my perspective, learning is a personal decision, it's a personal journey and it's even more difficult when you're an adult.

Provide the right support
While it's important to promote self-learning, it's also important to offer learning support, not just online support but support in real-life examples. It's important to create new spaces where people can learn from others. Of course it's important you promote the first step forward, so examples are very important to share good practices and success cases. If you want people to adopt new systems or new ways of working it's mandatory to shut down the old practices. This is the most difficult part; it's much easier to launch a new project than to kill an old one.

Create shared spaces
Building competencies is one part of the road map to help companies move to a new normality. It will be a corporate journey that will affect all aspects of the company–processes, infrastructure, procedures etc. It has to start by building infrastructure (spaces, processes, systems) to support new ways of behaviour. Then the next level is not only to adopt their offers to their customers but also to their employees. So the word 'experience' and 'design' are critical because companies are not just addressing infrastructure and systems, but they are designing new relationship models, new procedures, new processes linked to experiences and people. If you want the company to move fast, you will need employees to generate new content quickly. However this becomes difficult to monitor so instead of needing to review every single thing, you need to create rules of what is acceptable content. This is what I mean by 'creating shared spaces', which is about creating a playground for others to play in but with an agreed set of rules, and a common vision.

What is the future model of organisations?
The future of the organisations will be based on creating value in an ecosystem and being able to move from one opportunity to another. We have to be flexible and adapt to new ways of doing things together, similar to how a flock of bird moves. Understanding people and technology are critical. Not only do they help maintain a competitive advantage, they help our companies grow, expand, diversify and discover new businesses beyond the conventional boundaries of their industries.

Authors' biographies

JOYCE YEE, PhD is a design researcher, author and designer. She is currently an Associate Professor at the UK's Northumbria University's Design School teaching interaction, service and innovation approaches across undergraduate and postgraduate levels. In 2013 she published 'Design Transitions', a book about how design practices are changing. She has over 15 years working experience in a wide range of academic and professional environments as a user researcher and as an interaction and service designer. Joyce's research is focused on the value, role and impact of design in organisational contexts.

EMMA JEFFERIES, PhD is a business coach and service designer, and is the co-author of the book 'Design Transitions' (BIS, 2013). She holds a PhD in design and is part of the Design Thinkers Group. Her natural habitat is out in the wild, working with global innovation teams, consultancies and governments in South America, Asia, the Middle East and Europe. Emma helps organisations become more productive by facilitating people in working together to reveal a core purpose, daily practices and a vision. She is currently working with a UK government department creating empathy practices.

KAMIL MICHLEWSKI, PhD is a brand strategy and innovation consultant, author and speaker. Kamil was awarded a PhD on the subject of design and business in 2006. Over the last 10 years he has been helping blue-chip clients such as Electronic Arts, Marriott, Nestle, Sony, and Visa with challenges ranging from category growth and brand positioning to human-centred innovation. In 2015 Kamil published a book titled 'Design Attitude', which explores the nuances of design in the context of innovation and organisational cultures.